QGIS FOR HYDROLOGICAL APPLICATIONS

RECIPES FOR CATCHMENT HYDROLOGY
AND WATER MANAGEMENT

HANS VAN DER KWAST
KURT MENKE

PRESS

Credits & Copyright

QGIS for Hydrological Applications

Recipes for Catchment Hydrology and Water Management

by Hans van der Kwast and Kurt Menke

Published by Locate Press LLC

Direct permission requests to info@locatepress.com or mail:
Locate Press LLC, PO Box 671897, Chugiak, AK, USA, 99567-1897

Editor   Gary Sherman
Interior Design   Based on Memoir-LATEXdocument class
Publisher Website   http://locatepress.com
Book Website   http://locatepress.com/hyd

Contents

Foreword

Delft, July 2019

As an introduction to this book, Kurt and Hans approached me to write a foreword. Honored as I was when asked, I was very happy to say yes (it took around 0.2 seconds to make that decision). In the meantime I was thinking how I could best introduce this book to an audience that I don't yet know, and is probably much better educated in the field of Hydrology then I'll ever be. I could tell you more of the joys of working with GIS, or more specifically the joys of QGIS, because that is definitely more my cup of tea. And if there's one thing I'm absolutely positive about, it is that QGIS has a great tool set for a lot of use cases. But then, we have a book to consider here, and some things about the authors that they cannot in good conscience say about themselves. Well, that's what a foreword is for.

There are a number of books written on using QGIS. Most of them are generic in approach, teaching the "how to..." based on some case that might, or might not be helpful in other parts of the world. For example, editing of geometry is typically (not only in QGIS books) explained by making parcel maps. Now one thing you never do in the Netherlands is make your own parcel maps. Can someone explain to me why in the world you would want to make one? Don't you have a cadastre service that provides these? I wouldn't want to mess up a perfectly good cadastral map going around editing it myself, even if I do follow the rules of topological editing. But apparently a good centralized land registration system is not exactly common practice in the world...and thus, not all cases are applicable everywhere.

The field of Hydrology seems to be more internationally standardized. Water tends to flow down, it doesn't really matter where on the planet you are. It follows fundamental rules. Catchments can be calculated everywhere based on these rules. That makes the hydrological case an excellent one for examples in books on how to use QGIS. However, catchment delineation might be just a tiny bit specific for users who are not into hydrology.

This book doesn't have that problem, and it will help you discover QGIS for specific hydrological uses. It packs a lot of QGIS knowledge into this field, and in doing so, opens up the world for hydrologists to use the geospatial functionality of QGIS as a helpful tool.

All this practical knowledge comes from two great GIS educators, Kurt Menke and Hans van der Kwast. Kurt is known for some really quirky uses of QGIS—using the TimeManager plugin to make animated maps of where he has been teaching QGIS in recent times, for example. And sharing these on Twitter. Talk the talk, walk the walk...a typical Kurt-way of using QGIS. And of course he's much better known for his (other) books on QGIS as well. Apart from his books, doing an actual workshop with Kurt is a very enjoyable experience—he shows great interest in what people do (either with or without QGIS), and is ever happy to help. And his great love for QGIS makes sure that he's always the one showing a lot of practical uses for new features as soon as they're introduced in QGIS.

Due to events that happened in our mutual past, Kurt will visit the Netherlands only after promising him that there are no Belgian punk bands playing in the bar that we visit after work. We didn't say anything about said bands playing during work though...

I've known Hans a little longer, as he regularly organizes Missing Maps events at IHE in Delft, where he always invites me to help out. Normally, hardly any help is required after the first twenty minutes or so, so I guess he just invites me in to show off his excellent students. Which is okay, really. We tend to get into nice conversations, and being around international GIS-students in a more informal atmosphere is always an enjoyable way of exchanging knowledge and understanding.

So why are his students excellent, why does it seem that their understanding of geospatial issues all comes naturally? Partly of course because of their background. People who study at IHE don't just attend a class, they know what they want to learn and travel across the world to gain specific knowledge that IHE provides. So yes, maybe they're all just naturals with focus. However, that would be grossly neglecting the fact that they have an excellent educator. I know no other examples of a teaching staff member that a) single handedly set up a GIS curriculum for a specific field (Hydrology in this case), b) based it on open source software, c) shared his teaching resources online and d) set up e-learning, thus making it possible for hydrologists across the world to learn and use all necessary tools without having to worry about software licences and travel arrangements, or other practical problems one normally has to deal with.

The enthusiasm and energy Hans and Kurt have in teaching geospatial science and the use of QGIS to people across the globe is invigorating.

Now I'm really happy with this book, as it focuses on the things we use in everyday QGIS life. Every chapter starts with an explanation of a specific case that is common to the field of work. From there, the reader is helped step-by-step through the software. This combination of why and how makes it the excellent workbook for training, both in classrooms and for self-study.

Erik Meerburg
Founder of Geo Academie*
CEO LandGoed IT

The Dutch one, not the one in the U.S.

About This Book

When I started teaching GIS at IHE Delft Institute for Water Education in 2012, there were not many courses that used open source software. Given that our MSc participants are mostly from the Global South and can't afford expensive licenses, I wanted to change that for my GIS classes. QGIS was the logical alternative, which has all the features my students need for their work in hydrology and water management. In the recent versions, new features have been included that go far beyond what commercial GIS desktop software offers. In 2013 I started teaching QGIS in most of our MSc programmes and in short courses. In 2015 I had a great opportunity to develop new course materials with Jan Hoogendoorn (Vitens) for several trainings for the National Water and Sewerage Corporation (NWSC) in Uganda, funded by Vitens Evides International (VEI) and the IHE Delft Partnership Programme for Water and Development (DUPC). Jonne Kleijer also contributed to the materials during the trainings in Kampala and Lira.

At IHE Delft we had also started our OpenCourseWare platform in 2015. After the trainings in Uganda we agreed with the donors and trainers to make the course materials available as OpenCourseWare with a CC BY-NC license (`http://www.gisopencourseware.org`). This was an important step enabling many people to learn about QGIS for hydrological applications, even when they were not able to come to IHE Delft for our short courses or MSc modules. The course materials were completed with a YouTube channel (`https://www.youtube.com/c/HansvanderKwast`) with videos of the lectures and exercises. In the years that followed I regularly updated the materials following the QGIS Long Term Release (LTR) versions. Many MSc students at IHE Delft inspired me for improving the course materials and adding more instruction videos.

In August 2017 I joined a QGIS user conference and hackfest for the first time. This one was organised by Lene Fischer at Skovskolen Forest and Landscape College of the University of Copenhagen in Nødebo (Denmark). It was very inspiring to meet developers of QGIS and to learn about this open source community. Raymond Nijssen introduced me to different ways to contribute to QGIS. It was also here where I met Kurt Menke for the first time. Together with Tim Sutton we worked on the QGIS certification programme and its platform (`https://qgis.org/en/site/getinvolved/certification.html`). Since Nødebo I'm part of this great community.

During the short course on QGIS in September 2017, Erik Meerburg (Geo Academie) and I issued the first QGIS certificates. Since then IHE Delft has issued around 200 certificates, which is a great way for us to contribute to the further development of QGIS.

In January 2018 I was happy to host the first Dutch QGIS User Group Meeting at IHE Delft, organised in cooperation with Geo Academie. The tracks in Dutch and English attracted participants from diverse backgrounds.

For the short course on QGIS in 2018 I invited Kurt Menke as a guest lecturer to IHE Delft. It was inspiring for our participants to learn about the many new features of QGIS 3.x and how to become part of the community. It was then that the idea for this book was born.

During the QGIS user conference and hackfest in A Coruña (Spain) in March 2019 I followed Kurt's workshop and got inspired to develop a plugin for styling of flow direction maps using the functionality of the Crayfish plugin for styling mesh layers. Together with Radek Pasiok and Peter Petrik from Lutra Consulting we developed a processing plugin for this during the hackfest. It's available in the Crayfish plugin and works with QGIS 3.6 and later.

For the short course on QGIS for Hydrological Applications in September 2019 I've invited Kurt again. We will together launch this book and hope that many more GIS professionals get interested in using QGIS for hydrological applications and become part of the QGIS community.

Last, but not least, it is important to mention that by purchasing this book you are supporting students to attend FOSS4G and QGIS events. In this way I hope to contribute to increasing the diversity in the open source GIS community.

IHE Delft Institute for Water Education

IHE Delft Institute for Water Education (http:/www.un-ihe.org) is the largest international graduate water education facility in the world and is based in Delft, the Netherlands. The Institute confers fully accredited MSc degrees, and PhD degrees in collaboration with partner universities. Since 1957 the Institute has provided water education and training to 23,000 professionals from over 190 countries, the vast majority from Africa, Asia, and Latin America. Also, numerous research and institutional strengthening projects are carried out in partnership to strengthen capacity in the water sector worldwide. Through our overarching work on capacity development, IHE Delft aims to make a tangible contribution to achieving all Sustainable Development Goals in which water is key.

Since 2013 QGIS is the default in most MSc programmes. From October 2019 all MSc students will learn the concepts of GIS with QGIS in the first modules. Besides GIS education in the MSc modules, IHE Delft offers short courses, online courses, OpenCourseWare and tailor made training where the use of open source software (QGIS, Python) and open data is key.

Who This Book is For

The book is designed for professionals active in hydrology and water management, especially those involved in using simulation models for water management and performing GIS analysis. However, the book is also suitable for beginners in QGIS who want to learn GIS concepts and QGIS features in a problem based learning manner. The book can be used with an instructor in the classroom, or as independent study for beginners and experts. Each chapter introduces a case and concludes with styling recipes introducing the many features that QGIS offers. These concluding styling sections provide a solid foundation in the robust cartographic capabilities found within QGIS. Topics such as inverted polygon shapeburst fills, advanced label settings, and blending modes are covered.

The first chapter covers georeferencing and digitizing of vectors, in the second chapter the reader learns how to import spreadsheets, join and edit attribute tables, and interpolate point data to raster. Chapter 3 covers map algebra for spatial analysis. Different raster data types (Boolean, discrete, and continuous) are used in calculations, such as proximity analysis, reclassification, and Boolean logic. The Raster Calculator and functions in the Processing Toolbox are introduced.

The most common use of GIS for professionals in hydrology and water management is covered in Chapter 4 where the reader is guided through the procedure to perform stream and catchment delineation. Chapter 5 proceeds with showing how to add open data from web map services to the delineated catchment. Chapter 6 was inspired by my MSc thesis students from last year who needed percentages of land cover for each subcatchment. While going through the steps this chapter will also introduce vector geoprocessing functions such as intersect and dissolve. The chapter concludes with using the Data Plotly plugin for making pie charts. The book concludes with designing a well crafted map in the print composer.

The Data

The data for this book are available for download at `http://locatepress.com/qgis_hydrological` (178 MB). All the data are in the public domain and most of it can be downloaded from open data portals that are introduced in this book. The book also contains links to videos on theory and practice with QGIS.

About the Authors

Hans van der Kwast is senior lecturer at IHE Delft. He received a Master's degree in Physical Geography at Utrecht University in the Netherlands in 2002 with a specialization in GIS and Remote Sensing. In 2002 he was appointed at the Faculty of Geosciences of Utrecht University as a junior lecturer in GIS. In 2009 he defended his PhD at Utrecht University on the integration of remote sensing in spatial dynamic modelling of soil moisture using data-assimilation techniques implemented in the PCRaster Python framework. From 2007 to 2012 he worked at the Flemish Institute for Technological Research (VITO) where he was appointed as a researcher in spatial dynamic environmental modelling. He participated in projects related to water quality modelling, land-use change modelling, and the use of remote sensing for urban applications. As a guest lecturer he supervised BSc students in a hydrological and geomorphological fieldwork as part of the curriculum Earth Sciences at Utrecht University. Since April 2012 he works at IHE Delft. In his teaching he actively promotes the use of Free and Open Source Software (FOSS) by young professionals from the Global South. He is also involved in many research and capacity development projects related to GIS, remote sensing, spatial data infrastructures, Python, open data, and citizen science.

A former archaeologist, Kurt Menke is a geospatial generalist based out of Albuquerque, New Mexico, USA. He received a Master's degree in Geography from the University of New Mexico in 2000. That same year he founded Bird's Eye View (`https://www.birdseyeviewgis.com/`) to apply his expertise with GIS technology towards solving the world's mounting ecological, economic, and social issues. His main areas of focus are public health, conservation, and education. Kurt Menke has a broad skill set. He is a spatial analyst, cartographer, web map developer, trainer/teacher, and author.

Kurt has a long history using QGIS. He first downloaded it in 2005 when it was at version 0.7 Seamus. He is an open source GIS authority, having co-authored three editions of Mastering QGIS for Packt Publishing and authoring *Discover QGIS 3.x* (`https://locatepress.com/dq3`) through Locate Press (May 2019). He can frequently be found speaking at FOSS4G and QGIS conferences. In 2015 he became an OsGeo Charter Member. He is an experienced FOSS4G educator and is a co-author of the GeoAcademy (the U.S. version - `https://foss4geo.wordpress.com/`). He is now a QGIS Certified Instructor. His offerings range from a semester long Intro to FOSS4G course he originally developed in 2009, to short courses and professional workshops. In partnership with the U.S. National Library of Medicine (NLM) he created a program named Community Health Maps (`https://communityhealthmaps.nlm.nih.gov/`). This program aims to empower underserved and minority populations with open source mapping technology. Kurt has trained over 800 public health workers in this workflow over the last five years. In 2015 he was awarded the Global Educator of the Year Team Award by GeoForAll as part of the GeoAcademy team.

Acknowledgments

Hans would like to thank IHE Delft and his colleagues for supporting him in developing high quality curricula with open source GIS for their MSc students. He's very grateful to have the chance every year to introduce around 200 MSc, PhD, and short course participants from the Global South to open source software. He would also like to thank these participants who inspire him to keep up the good work and teach the state of the art. Hopefully we can make a change and get rid of the vendor lock-in in many countries.

In addition, he would like to thank the donors and project partners who supported the development of course materials with QGIS, enabling their beneficiaries to learn open source alternatives.

He would like to thank Kurt Menke for being his co-author. It's great to work with Kurt. Without him the layer styling sections and map design chapter would not have been of exceptionally high quality.

Kurt is grateful to Hans van der Kwast for inviting him to partner on this book. After being invited as a guest lecturer at IHE Delft in 2018—teaching this same material—it's exciting to see it come to fruition in book form. He would like to thank the QGIS community for constantly sharing knowledge and inspiring him to use new techniques in the ever evolving QGIS project.

We would both like to thank Erik Meerburg for his delightful Foreword.

Finally, we would like to thank editor and publisher Gary Sherman, the founder of QGIS and owner of Locate Press publishing this book. He is always a keystroke away, helping us out with questions about LaTEX, RST, and git. He also helped edit this book.

A note from Hans: I'm writing this section at the same location where I did my PhD research: the rural community of Sehoul in Morocco. A nice place for a writer's retreat but moreover a place where I keep my feet on the ground to keep in my mind why the skills in this book are so important: to improve catchment management and protect vulnerable areas and people from floods, droughts and environmental problems.

1. Preparing Data from Hardcopy Maps

1.1 Introduction

In order to use hardcopy maps in a GIS, they need to be scanned and georeferenced. Georeferencing is also needed for raw remote sensing images, such as aerial photographs and satellite images.

For the best result, choose a map sheet that is clean and does not have too many folds. Use a scanner that is large enough to scan the whole map. The resolution of the scanner should be high enough (e.g. 1200 dpi) to have sufficient detail in the resulting raster maps.

For georeferencing we need to link locations on the scanned image to coordinates. There are two methods:

- Collect ground control points (GCPs) at locations that are clearly visible in the image, such as bridges and junctions.
- If the hardcopy map contains a coordinate grid, use the printed grid as a reference. Make sure that you know the projection of the grid, which is usually stated on the map.

After this chapter you will be able to:

- find the projection and *EPSG* code of a map
- install plugins
- *georeference* a scanned map using *GCPs* from a grid
- use the *coordinate capture tool*
- use online layers from the *QuickMapServices* plugin
- digitize points, lines, and polygons and add attributes
- use the *snapping toolbar*
- *dissolve* features
- store data in a *GeoPackage*
- style and label vectors

Figure 1.1, on the following page shows the workflow of this chapter.

In this chapter we will use a scanned map of Mount Marcy (USGS, 1979), `Mount_Marcy_New_York_USGS_topo_map_1979.JPG`, which we will georeference with the coordinate grid printed on the map. You can download the file from `http://locatepress.com/qgis_hydrological`. Save it to a local folder.

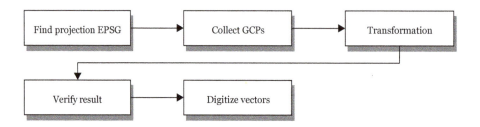

Figure 1.1: Workflow of this chapter

For this chapter the following videos are recommended as theoretical preparation: "Introduction to GIS, "GIS Vector data model", and "Map projections in GIS". Videos can be found at the YouTube channel http://www.youtube.com/c/hansvanderkwast.

1.2   Choosing the Projection

Have a look at the scanned map and try to find the projection used. You can use any image viewing software for this.

- Which projection was used?

- Look for the EPSG code at http://www.spatialreference.org and write it down.

1.3   Enable the Georeferencer GDAL Plugin

The Georeferencer GDAL plugin is a core plugin, which means it is already installed with QGIS. In order to use it, you need to activate it.

1. In the main menu choose Plugins | Manage and Install Plugins.

2. Search for Georeferencer GDAL and check the box (figure 1.2). Click *Close* to close the dialogue.

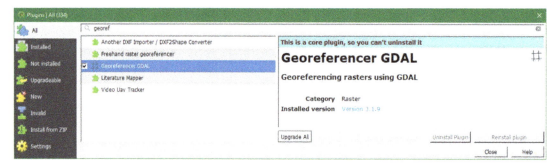

Figure 1.2: Installing the Georeferencer GDAL plugin

1.4   Importing the Scanned Map into the Georeference GDAL Plugin

1. From the main menu choose `Raster | Georeferencer`

2. Click the *Open Raster* button .

3. Browse to the `Mount_Marcy_New_York_USGS_topo_map_1979.JPG` file. A window will open where you have to specify the *Coordinate Reference System (CRS)* of this input map. It does not yet have a CRS, so you can click *Cancel*.

1.5   Setting the Transformation Parameters

First we have to set the transformation settings.

1. In the georeferencer menu choose `Settings | Transformation Settings`; here you can choose:

- Different *transformation types*. A simple linear transformation can be used if the map is not much deformed. The other ones can be tried when more deformation exists. More information about the transformation types can be found in the plugin documentation `https://bit.ly/2LMGKGs`. We will start with a linear transformation.

- *Resampling method*: if you need the pixel values in further calculations, it is best to choose the nearest neighbor option. This resampling method will preserve as much as possible the original pixel values by choosing the nearest one. Visually, however, this method results in a "blocky" map. If the purpose is only for visual use, for example as a backdrop for digitization of vector layers, then it is better to choose another resampling method. Here we will use the cubic method, which uses the average of the 16 nearest pixels.

- Target SRS: here choose the code that you have noted before; EPSG:26718. You can choose it by clicking 🌐 and typing `26718` in the filter field.

2. Browse to the folder where you want to save the georeferenced map. The tool automatically adds `_modified` to the file name. So in our case the georeferenced file will be named `Mount_Marcy_New_York_USGS_topo_map_1979_modified.tif`. Keep the other settings on default and check the box *Load in QGIS when done*.

The dialogue should look like figure 1.3, on the following page.

1.6   Adding Ground Control Points (GCPs)

In order to link the file coordinates to real world coordinates, we need to indicate Ground Control Points (GCPs) with known coordinates. We can derive these coordinates in different ways:

- The easiest way is to use the coordinate grid on the scanned map if this is available and if it is in a known projection. We click on a node in the grid and type the corresponding X and Y coordinates in the dialogue.

- Using a reference map in the QGIS map canvas that has already been georeferenced. In this way we can obtain the right coordinates by clicking on the reference map.

- Using GCPs that were measured in the field using a GPS.

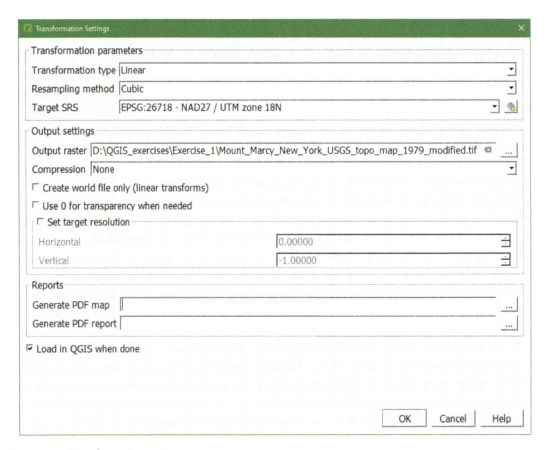

Figure 1.3: Transformation settings

Here we will use the coordinate grid that is printed on the map.

1. Zoom in on the node with coordinate 581000 East and 4885000 North.

2. Click the *Add Point button* to add a GCP.

3. Enter the map coordinates in the pop-up window (figure 1.4).

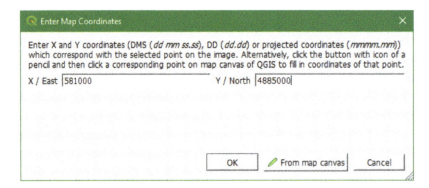

Figure 1.4: Enter Map Coordinates

> If you have a reference map in the QGIS map canvas, you can use the *From map canvas* button to capture the coordinate and perform an image to image georeferencing.

4. Press *OK*.

Now your screen should look like figure 1.5.

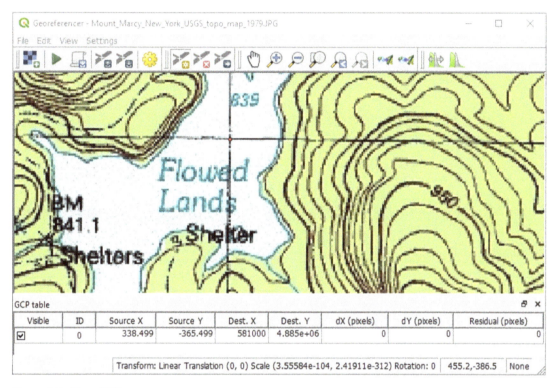

Figure 1.5: First GCP

The red dot is the location that you have referenced. In the table below the map, you can see the *Source X* and *Source Y* coordinates. These are the unreferenced file coordinates. Their values depend on which pixel you clicked for placing the GCP, so it can differ from the screenshot in figure 1.5. *Dest. X* and *Dest. Y* show the real world coordinates that you have linked to this location. The other fields of the table have to do with estimated accuracy and will be filled in after adding more points.

5. Let's choose a second GCP in the upper right corner of the map and proceed in a similar way as with the first GCP. Your screen should look like figure 1.6, on the next page.

You can see that some error statistics have been calculated. With only two points this does not make much sense. The minimum amount of GCPs for a linear transformation should be three. With more GCPs that are well distributed over the map, the transformation will be more accurate.

6. Add in a similar way a GCP in the lower left and the lower right corner of the map. If you made a mistake you can remove the GCP by using the *Delete point* button ✂. Your screen should look like figure 1.7, on the following page.

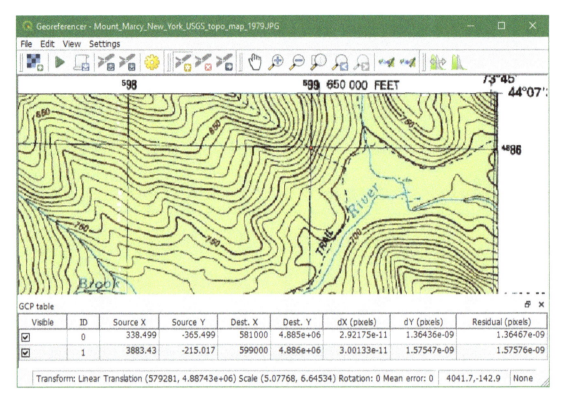

Figure 1.6: Second GCP

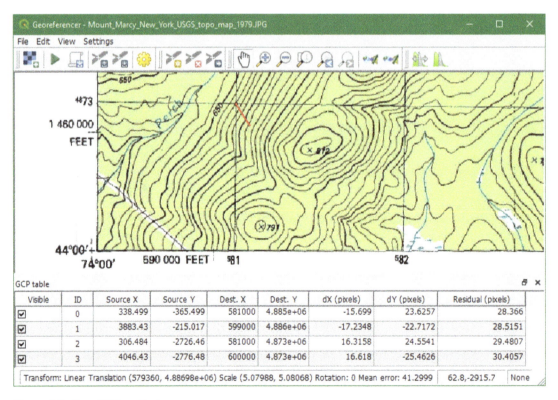

Figure 1.7: Four GCPs added

1.7 Reduce Errors and Perform the Transformation

At the bottom of the screen you can see the estimated mean error (41.2999 pixels in our case). The error is also visualized at the GCPs using a red line. Obviously your values will not be exactly the same.

There are two ways to reduce the error:

- Use the *Move GCP Point* button to place the GCPs really at the nodes where the grid lines cross. You need to zoom in to select the right pixel. Note that the mean error is not automatically updated. You need to change the transformation type to something else and then back.

- If the first option doesn't reduce the error, we can change the transformation type. If we change to another transformation type in the transformation settings, the error values will be recalculated.

In this exercise we'll apply the second option.

1. In the menu, go again to `Settings | Transformation Settings`

2. Now let's select a first order polynomial (*Polynomial 1*) instead of the linear transformation. Keep the rest as it was.

3. Click *OK* to return to the GCP table.

Now you can see that the mean error has been reduced to a fraction of a pixel, which is acceptable. If you don't see a mean error < 1 pixel, you will have to check the GCP locations and correct them.

> A mean error < 1 pixel cannot always be achieved. The decision to accept a certain accuracy depends on the purpose of the map.

4. Now we can start georeferencing using the button.

After some calculation time the georeferenced map appears in the QGIS Map Canvas.

5. You can close the *Georeferencing* plugin. It will ask if you want to save your GCPs. You can click *Discard* if you don't want to use them. If you save them, you can load them again in the Georeferencing plugin.

1.8 Verify the Georeferenced Map

In order to verify the result you can use the *Coordinate Capture* plugin.

1. If it is not activated yet you can do so by choosing `Plugins | Manage and Install Plugins` from the menu, then search for the *Coordinate Capture* plugin (figure 1.8, on the next page). Check the box to activate it and click *Close*.

A panel has appeared under the layers list.

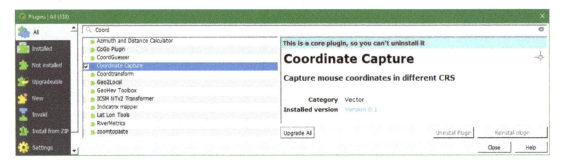

Figure 1.8: Install the Coordinate Capture plugin

2. Click on *Start capture*.

3. Click on a grid node in the map and the coordinates are displayed in the panel (figure 1.9).

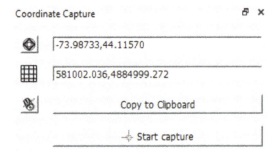

Figure 1.9: The Coordinate Capture panel

The first field shows the coordinates in the Geographic Reference System (Lat/Lon coordinates), the second field shows the coordinates in the projection of the project.

4. Read the coordinates from the side of the map and verify if they correspond with the coordinates in the *Coordinate Capture* panel.

Another way to verify the result is to use the web maps as a backdrop. The *QuickMapServices* plugin provides easy access to many web maps such as Google satellite and OpenStreetMap.

If not yet installed, install the *QuickMapServices* plugin in the same way we installed other plugins. After installing, choose Web | QuickMapServices plugin | Settings from the main menu. Next, choose the *More services* tab and click *Get contributed pack* (figure 1.10, on the next page). Click *Save* when the contributed pack is installed. This will give you access to many more useful web maps.

5. Go to the main menu and choose Web | QuickMapServices plugin and select Google | Google Terrain and then OSM | OSM Standard. Compare the georeferenced map with Google Terrain and OpenStreetMap.

To help with this comparison you will employ a *Blending Mode*. Blending modes determine how two layers interact visually. When a blending mode is applied to a layer it will be blended with the layer below.

6. Open the *Layer Styling panel* by clicking the 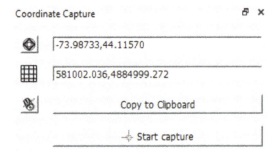 button. Set the target layer to *Mount_ Marcy_New_York_USGS_topo_map_1979_modified*.

Figure 1.10: QuickMapServices Settings

7. In the panel, choose *Blending mode Multiply* (figure 1.11).

> Blending modes allow for more elegant rendering between GIS layers. They can be much more powerful than simply adjusting layer Opacity. Blending modes allow for effects where the full intensity of an underlying layer is still visible through the layer above. There are thirteen blending modes available.

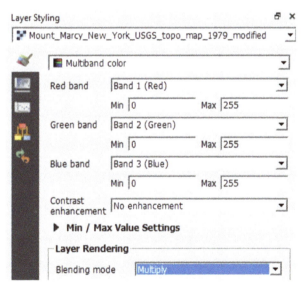

Figure 1.11: Setting the blending mode to multiply

This is a good time to save your QGIS project (.qgz file).

8. From the main menu, choose Project | Save as.

The project file contains references to all layers, styles, projections, extent and zoom level of the Map Canvas. Save your project regularly!

> Try to use the Georeferencer plugin to register the scanned map to a satellite image from the QuickMapServices plugin. In this way you can perform an image to image registration. Please make sure you use the right projection. Does the image to image registration give better results?

1.9   Digitizing Vector Layers from a Georeferenced Backdrop

Our georeferenced scanned map can now be used as a backdrop to digitize vector layers. Vectors can be points, (poly)lines or polygons. In this exercise we are going to digitize:

- Mountain tops as points
- Rivers as polylines
- Lakes as polygons

We will create these vector layers in a GeoPackage spatial database. In that way we have all layers together in one file, instead of using separate shapefiles.

Digitize Peaks

First we have to create an empty GeoPackage layer.

1. From the main menu select Layer | Create Layer | New GeoPackage Layer. You can also use the *New GeoPackage Layer* button, or the keyboard shortcut for your platform (e.g. Ctrl + Shift + N on Windows).

2. In the *New GeoPackage Layer* dialogue, first select the Database filename by clicking the ... button. Browse to the folder where you want to store the GeoPackage and save it to Mount_Marcy.gpkg. For *Table name* choose Peaks. For *Geometry type* choose *Point*. Make sure EPSG:26718 is chosen as the projection. Create a new field with the *Name*: Elevation, *Type*: Whole Number (integer) and click Add to Fields List (figure 1.12, on the next page).

3. Add a second new field with *Type*: Text named Name. Click *OK*.

The empty layer has now been added to your layers list (figure 1.13, on the facing page).

In order to start digitizing, you have to toggle to editing mode.

4. Click on the *Peaks* layer so it is selected. Click on to toggle to editing mode. You can also right-click on the layer and choose *Toggle Editing* from the context menu to place the layer into edit mode.

Now the buttons on the *Digitizing Toolbar* become active. A pencil icon also appears next to the layer in the Layers Panel indicating the layer is in edit mode.

5. On the topographical map navigate to a mountain. If a geodetic datum is present on top there will be an X and an elevation value.

6. When you have found one, zoom in and click the *Add Point Feature* button . The cursor changes to a crosshair. Move the mouse to the mountain top. Click on the mountain top.

Figure 1.12: New GeoPackage layer dialogue

Figure 1.13: Peaks added to the layers list in the layers panel

A dialogue with a form shows up. *fid* is the feature id that is automatically generated. It's a unique integer number for each feature. The other attributes that we have to fill in are Elevation and Name.

7. In this example we type 1501 and Mt Skylight for *Elevation* and *Name* respectively (figure 1.14, on the next page). If you map a peak without a labeled elevation determine the elevation based on the contour lines.

8. Repeat this step for a few other peaks.

If you made a mistake, you can use the *Vertex Tool* ![vertex tool icon] to move the feature or use one of the

select options (figure 1.15, on the following page) to select the point feature and click ![trash icon] to

Figure 1.14: Add Feature attribute of Peaks

delete the selected point feature. These buttons ⬅ ➡ can be used to undo/redo editing actions. Use 💾 to save the edits.

Figure 1.15: Select options

9. When done, click again on the ✏ button to toggle editing. If you didn't save edits yet, it will ask you to *Save* or *Discard*. With *Discard* you can always undo your edits until the last time it was saved.

10. You can check the attribute table of your new point vector layer by right-clicking on the layer name (*Peaks*) and selecting `Open Attribute Table` (figure 1.16, on the next page).

Now you can see the attributes that you have added and their *fid*, *Elevation*, and *Name* values (figure 1.17, on the facing page).

Digitize Rivers

Your next task is to digitize line features (rivers). The procedure is similar to creating a point layer.

11. In the *New GeoPackage Layer* dialogue first browse to the existing *Database,* `Mount_Marcy.gpkg`.

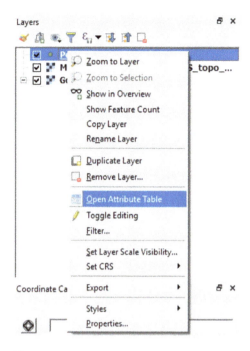

Figure 1.16: Open attribute table

Figure 1.17: Attribute table of the peaks layer

12. Give a new *Table name* of Rivers and a *Geometry type* of Line. As a new attribute we add Name with the type *Text data*. Don't forget to click the [Add to Fields List]. Check if the dialogue resembles figure 1.18, on the next page and click *OK*.

A *New GeoPackage Layer* window will open informing you that the file already exists.

13. Choose *Add New Layer* (figure 1.19, on the following page).

The *Rivers* layer is now added to the layers panel.

14. Toggle editing to start adding rivers. Click on the *Add Line Feature* button to add a new river feature. Zoom and pan on the map to find a stream to digitize.

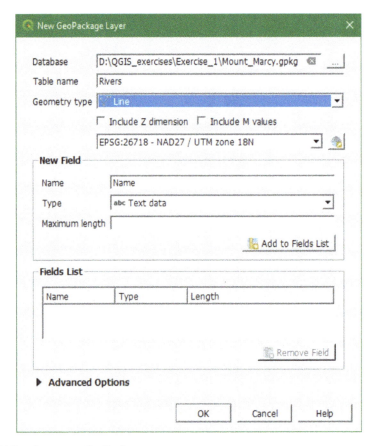

Figure 1.18: Add line feature to GeoPackage

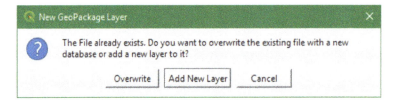

Figure 1.19: Add a new layer to the GeoPackage

Always start digitizing from the upstream to downstream. The direction will be stored in the layer. Always place a vertex when a tributary joins a larger stream. That's important when connecting the tributary later.

15. Click on the starting point of the line (node) and click when necessary to make a vertex.

You can use the zoom and pan buttons to trace the stream. You can use the spacebar to pan during digitizing. With Backspace you can delete the last node while digitizing.

16. After you place the end node of the line, right-click to complete the feature. Now you can enter the attribute data in the form (figure 1.20, on the next page).

Now you are going to digitize a tributary. First you have to set the snapping options.

Figure 1.20: Add line feature attributes

17. In the main menu choose View | Toolbars | Snapping Toolbar (alternatively right-click on a toolbar and choose Snapping Toolbar). Click  to enable snapping. Choose to snap to vertices of the active layer with a tolerance of 15 meters (figure 1.21).

Figure 1.21: Snapping toolbar

18. Now digitize the tributary from upstream to where it joins the higher order river. You will see that the line will snap to the node that you placed before on the main river.

If you want all tributaries to be one single feature, you need to *dissolve* the features.

19. When you have given all tributaries the same name, you can dissolve them by choosing Vector | Geoprocessing Tools | Dissolve from the menu. Use the ... button to choose Name as the *Dissolve field* and save the result in a GeoPackage layer with the name Rivers_dissolved. Click *Run* (figure 1.22).

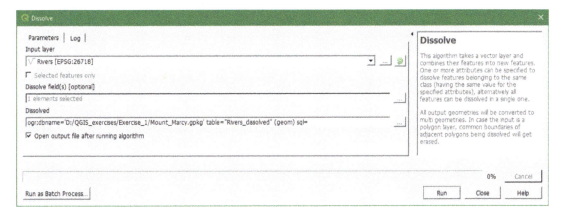

Figure 1.22: Dissolve

20. Check the attribute table of the result (figures 1.23 and 1.24, on the following page).

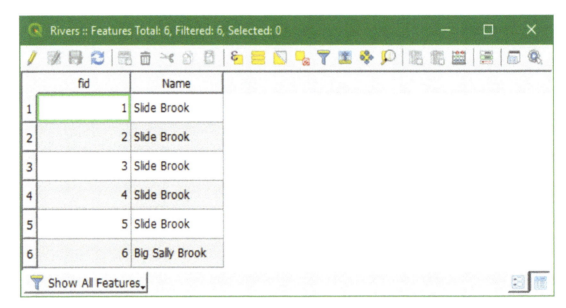

Figure 1.23: Attribute table before dissolving the tributaries

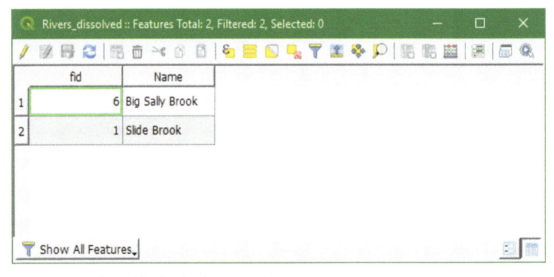

Figure 1.24: Attribute table after dissolving the tributaries

Digitize Lakes

Finally we are going to create a polygon vector layer for some lakes. Try to find out how to do this for yourself . It is very similar to the procedure for lines. The only difference is that the first node should be the same as the last node in order to close the polygon. Use the name of the lake as a text attribute.

Some lakes have islands that need to be removed from the lake features. For this purpose, enable the *Advanced Digitizing Toolbar* in a similar way as we enabled the *Snapping Toolbar* earlier. Click and digitize the island in the same way as you digitize a polygon. After finalizing the ring, it will be removed from the lake polygon.

1.10   Styling the Mountains, Rivers, and Lakes

In this final section you will style your peaks, rivers, and lakes.

Style Peaks

You will begin with the Peaks layer.

1. Open the *Layer Styling panel* by clicking the ✎ button. Set the target layer to *Peaks*.

By default they are styled using a Simple marker.

2. Select *Simple marker* and change the *Symbol layer type* to *SVG Marker*. Below you will find a section named *SVG Groups*. Here you can browse for SVG icons that are installed with QGIS. Find the *symbol* folder and click on it. Scroll down to find the *red marker* 📍 icon and select it. Change the *Width* and *Height* to 12 each. See figure 1.25.

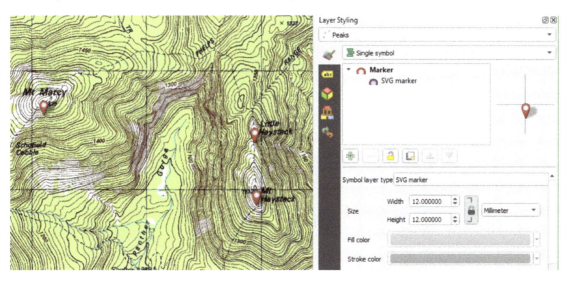

Figure 1.25: Peak SVG Symbols

Label Peaks

Next you will label the peaks.

3. Switch to the *Labels* tab **abc** of the *Layer Styling Panel*. Switch from *No Labels* to *Single labels*. Set the *Label with* option to the Name field. It is also possible to use multiple fields in a feature label by using an expression.

4. Click the *Expression* **ε** button to open the *Expression Dialog* window. Expand the *Fields and Values* section and add the Elevation field after the Name field. When combining text elements in an expression they need to be separated by the *String Concatenation* **||** operator.

5. Additionally, the *New line* **'\n'** operator can be used to wrap the new column onto a second line. However, it requires another *String Concatenation* operator after it. Set up an expression

like this: "Name" || '\n' || "Elevation". This is nice, but it may not be clear what the number represents and the labels will be easier to read with a thousands separator.

6. To add units of measure to the elevation, you can add the string *meters* after the value by appending || ' meters' to the existing expression. To accomplish the second enhancement you will use the format_number function. Use the search box to find the format_number function. Insert it right before the Elevation field. The help panel will show you the syntax for this function. It requires a number and a number of decimal places. The number will be the Elevation field and the number of places 0. This will simply format the data to a number and insert a thousands separator. See figure 1.26.

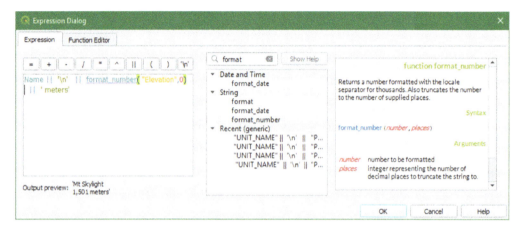

Figure 1.26: Peak Labeling Expression

7. To make the labels easier to read change the font *Style* to Bold. Switch to the *Label buffer* tab **abc** and check the *Draw text buffer* option. To give more separation between the labels and the feature icon switch to the *Label placement* tab and set the *Distance* to 2.

8. Finally click the *Automated placement settings* button to open the *Automated Placement Engine* window. Uncheck the box for *Allow truncated labels on edges of map* option. This will prevent labels from being cut off.

9. Click *OK* to dismiss and your labels should look like figure 1.27, on the next page.

Style Rivers

Next you will style the rivers.

10. Set the target layer in the *Layer Styling Panel* to *Rivers* and switch from the Labeling tab back to the Styling tab.

11. Select *Simple line*. Click the *Color* bar to open the *Select line color* panel.

12. Change the *Color* to an RGB value of 31|120|180.

13. Click the *Go back* button to return to the main symbology panel.

14. Increase the *Stroke width* to 0.86.

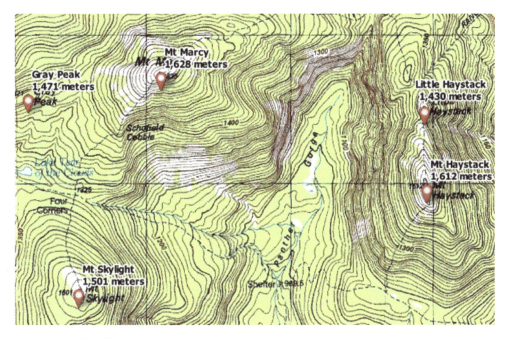

Figure 1.27: Peak Labeling

15. To label the rivers switch to the *Labels* tab ⬛abc.

16. Repeat the initial steps of labeling Peaks to label rivers with just the Name field.

17. Switch to the *Label placement* ✦ tab and choose a *Placement* style of *Curved*.

To make them more readable against the topo map you will apply a buffer.

18. Switch to the *Label buffer* tab abc and check *Draw text buffer* option.

You will set the color to the green background of the topo map.

19. Click the drop down arrow for the *Color* setting and choose *Pick color*. With the eye dropper cursor click on a place to select that green topo map background color.

20. Finally return to the *Text* tab abc and set the label *Color* to an RGB value of 31|38|180, the *Font* to Calibri, the *Size* to 11 and the *Style* to *italic*.

Your rivers should look like figure 1.28, on the following page.

Style Lakes

Next you will style the lakes. You will use a shapeburst fill which will allow you to color the lakes from light blue to dark blue.

21. Set the target layer in the *Layer Styling Panel* to *Lakes* and switch from the Labeling tab back to the Styling tab. Select *Simple line*.

22. Select the *Simple fill* styling component. Change the *Symbol layer type* to *Shapeburst fill*. Keep the default *Gradient colors* setting of *Two color*. Set the first color to an RGB value of

Figure 1.28: River Styling

185|239|255. Set the second color to an RGB value of 31|133|180.

23. Set the *Shading style* to *Set distance* with a value of 6. Increase the *Blur strength* to 12.

Finally you will add a simple line to represent the coastline of each lake.

24. Click the *Add symbol layer* button. Change the new *Simple fill* renderer to a *Symbol layer type* of *Outline simple line* with *Color* of 31|133|180.

25. Label the lakes with the the Name field. Set the label *Color* and RGB value of 225|255|255 (light blue), the *Font* to Calibri, the *Size* to 10 and the *Style* to *bold italic*.

26. Switch to the *Label placement* tab and change the *Placement* to *Horizontal (slow)*. Then switch to the label *Formatting* tab and enter a space as the *Wrap on character*. Set the *Alignment* to *Center*.

Your lakes should look like figure 1.29, on the next page.

The following videos explain the steps in this chapter and the image to image registration: "Georeferencing a scanned map and digitizing vectors in QGIS3" and "Image to image georeferencing in QGIS 3.4". Videos can be found at the YouTube channel http://www.youtube.com/c/hansvanderkwast.

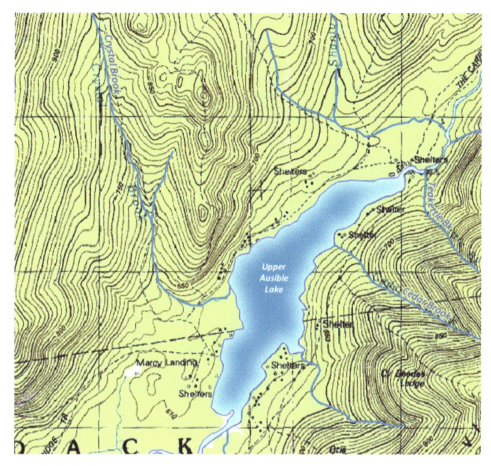

Figure 1.29: Lake Styling

2. Importing Tabular Data into QGIS

2.1 Introduction

Often we get data in tabular format, for example spreadsheets or CSV files. Sometimes the data comes in two tables—one with the coordinates and another one with the attributes you need for your analysis.

After this chapter you will be able to:

- *import tabular data* into a GIS
- save tables with geometry to GIS format
- *join attribute tables*
- *edit attribute tables*
- *interpolate* point data to raster
- style and label point vectors
- style continuous rasters

In this example we are going to import a table with the daily average temperature on September 1, 2013 at several meteorological stations in the Netherlands. The data was downloaded from the KNMI Data Centre (KNMI, the Royal Netherlands Meteorological Institute, `http://data.knmi.nl`), but reformatted for the purpose of this exercise.

Figure 2.1 shows the workflow of this chapter.

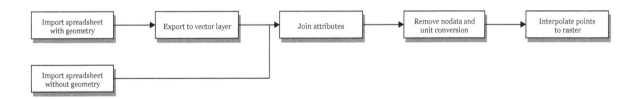

Figure 2.1: Workflow of this chapter

In this exercise we'll use the following data:

- `KNMI_20130901_tday.xls`: table with average daily temperatures for different stations
- `KNMI_stations.xls`: table with station number and coordinates of the location of the stations

This data can be downloaded from `http://locatepress.com/qgis_hydrological`.

> For this chapter the following video is recommended as theoretical preparation: "Raster data model in GIS". Videos can be found at the YouTube channel http://www.youtube.com/c/hansvanderkwast.

2.2 Check the Spreadsheets

It is good practice to check the contents of spreadsheets or delimited text files before importing them into any software. For CSV files, for example, it is important to know the column separator.

Open the files `KNMI_20130901_tday.xls` and `KNMI_stations.xls` in a spreadsheet program and check the contents.

- Which file contains coordinates? Is there a way to link both files? How could we do that?

2.3 Import Spreadsheets

There are different ways in QGIS to import tabular data:

- `Layer | Add Layer | Add delimited text layer`. This is the standard importer that allows us to import delimited text files.

- `Layer | Add Layer | Add spreadsheet layer`. This tool can load spreadsheet files (`*.ods`, `*.xls`, `*.xlsx`) as a layer with options to use the first line as a header, ignore rows, and load geometry from x and y fields.

Here we'll use the second option, for which we need to install the *Spreadsheet Layers plugin*.

1. Start QGIS Desktop. Make sure you start a new project rather than continuing the previous one.

2. In the main menu go to `Plugins | Manage and Install Plugins` and check if the *Spreadsheet Layers plugin* is installed (figure 2.2, on the next page). If not, install it now.

3. Now choose `Layer | Add Layer | Add spreadsheet layer` from the main menu (figure 2.3, on the facing page).

4. In the dialogue browse to the file with the locations of the meteorological stations (`KNMI_stations.xls`).

5. Fill in the dialogue as in figure 2.4, on page 38. Make sure the right *Geometry Fields* and *Reference system* are chosen. Also indicate the data types correctly for the fields. For example, `STN` are station numbers and should be imported as *Integer*, while `ALT(m)` should be imported as *Real* numbers.

Once you click OK, a map with the meteorological stations is displayed.

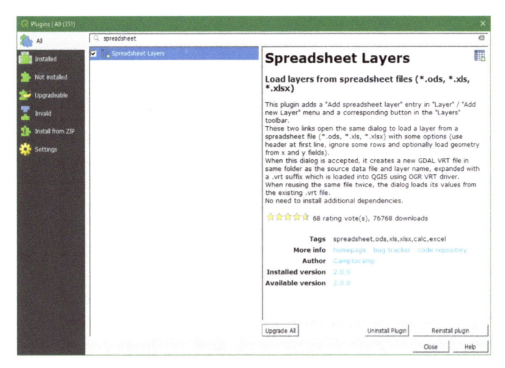

Figure 2.2: Install the Spreadsheet Layers plugin

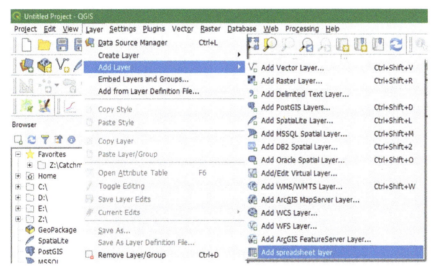

Figure 2.3: Add spreadsheet layer

6. If you don't see the map, you probably need to zoom to the extent of the map. Right-click on the layer name (*KNMI stations table*) and choose Zoom to layer.

7. Add the table with the temperature data in the same way. Because there is no geometry (coordinates) in the table, we should uncheck the box (figure 2.5, on page 39).

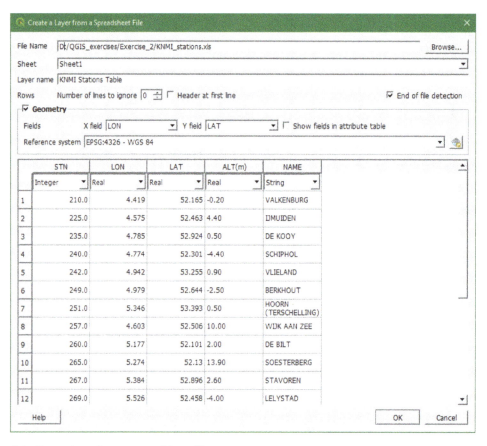

Figure 2.4: Create layer from a spreadsheet file

2.4   Convert Spreadsheet to Vector Layer

Now the layer *KNMI stations table* is saved as a temporary virtual layer. For further processing we first need to convert it to a GIS vector format. In this case we'll convert it to a shapefile.

1. Right-click on *KNMI stations table* and choose Export | Save features as

2. In the dialogue, use the ⸬⸬⸬ button to browse to the right folder to save the file as KNMI_ stations.shp. In order to change the projection to the local Dutch projection, choose Amersfoort / RD New for the *CRS* by clicking on the 🌐 button. Tip: use the Filter field to lookup EPSG code 28992 (figure 2.6, on page 40).

> Here you see the advantage of using EPSG codes: they provide standard codes for each projection. It is useful to determine the EPSG code of the projection you want to use for your project.

3. Click *OK*. Now the dialogue looks like figure 2.7, on page 40 (also check the box *Add saved file to map* and make sure that ESRI Shapefile is chosen as the *Format*).

4. Click *OK* to proceed.

5. Remove the *KNMI Stations Table* from the layers list by right-clicking and selecting Remove Layer. Click *OK* to confirm. Be sure to remove the right one. If you hover your mouse over the

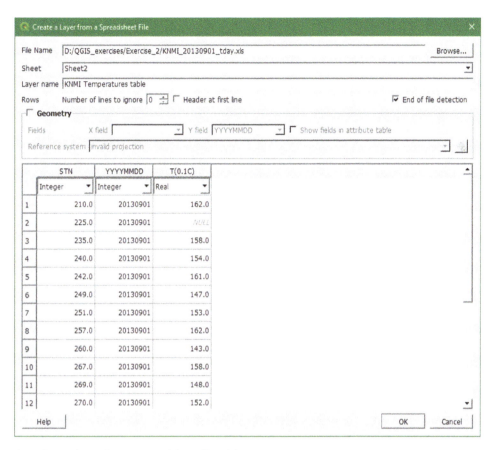

Figure 2.5: Create layer from a spreadsheet file without geometry

layer item it will show the file name. With Remove Layer you only remove it from the layers list, the file will still be on your hard disk.

Although the knmi_stations.shp layer is in the EPSG:28992 projection (Amersfoort / RD New), the QGIS Map Canvas still uses the EPSG:4326 projection (lat/lon WGS 84) and has reprojected knmi_stations.shp on the fly for visualisation. In order to visualise all layers in EPSG 28992 we have to change the QGIS Project properties.

6. In the main menu choose Project | Properties.

7. Choose the *Coordinate Reference System (CRS)* tab (figure 2.8, on page 41).

8. From the list of recently used coordinate reference systems, choose EPSG:28992 and click *OK*.

> Another way to set the project CRS to that of the layer, is to right-click on the layer and choose *Set CRS | Set Project CRS from Layer* from the context menu.

Note that the projection of the project is indicated in the lower right of the screen: 🌐 EPSG:28992 . You can always check there to see if the EPSG code is okay. This is also a button you can click to go to the Project Properties | CRS tab and change the on-the-fly reprojection.

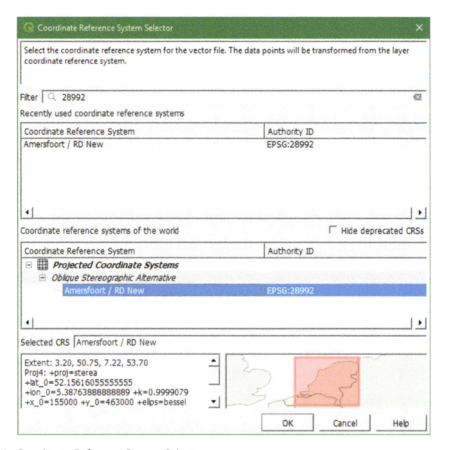

Figure 2.6: Coordinate Reference System Selector

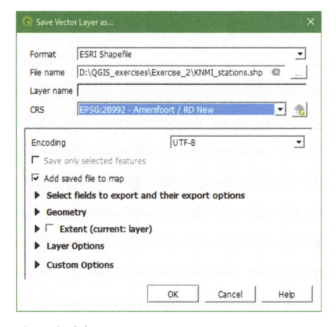

Figure 2.7: Save Vector Layer As dialogue

2.5 Join Attribute Tables

The locations of the meteorological stations and the temperature data are still in separate tables. For further analysis we need to combine them into one vector layer. In GIS terms this is called

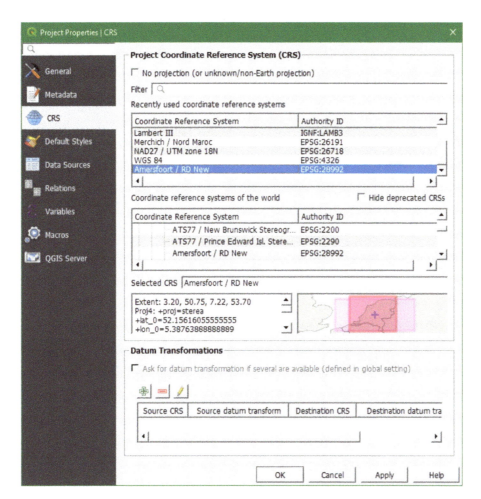

Figure 2.8: Change the on-the-fly reprojection of the project

a *join* operation. We can only join tables if they have a column in common.

1. Check the attribute table of *KNMI_stations* (right-click on KNMI_stations and choose Open Attribute Table) and in the same way, check the *KNMI Temperatures Table*.

- Which column do both attribute tables have in common?

After determining which column both tables have in common we can join the data of *KNMI Temperatures Table* to the attributes of our shapefile KNMI_stations.shp.

2. First close the attribute tables.

3. Next, right-click on *KNMI_stations* and choose Properties.

4. In the dialogue choose the button *Joins*

5. Click the button check if the dialogue looks like figure 2.9, on the following page.

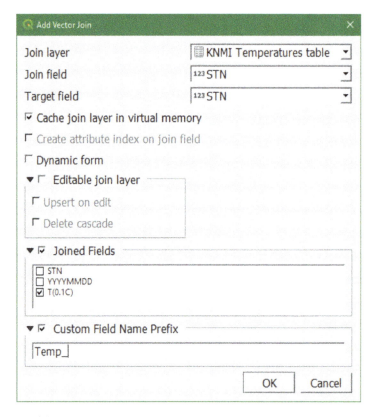

Figure 2.9: Add vector join

The common field is STN (the station number). We join only the temperature field and give the column the prefix Temp_.

6. Click *OK*.

7. Click *OK* to perform the Join operation.

Now check again the attribute table of *KNMI_stations*.

- What happened?

2.6   Edit the Joined Attribute Table

The joined attribute table needs two corrections: (1) features with missing data need to be removed, and (2) the temperatures have to be converted to the right units.

1. Click on row numbers with NULL or no values for temperature, while keeping the Ctrl key pressed.

Now the attribute table looks like figure 2.10, on the next page.

2. In the attribute table click on ✏ above the table to toggle editing mode.

Figure 2.10: Select features with missing data

3. Click the 🗑 icon (in the toolbar above the attribute table) to remove the two features with missing data, then save the attribute table by clicking 💾.

The only problem now is that the temperatures in the table are in 0.1 °C. We need to convert the values to °C.

4. Click the *New field* button 📇 to add a new column to the table. Fill in the *Add Field* dialogue as shown in figure 2.11.

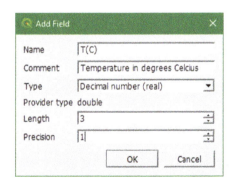

Figure 2.11: Add field dialogue

Length is the number of digits, *Precision* is the number of decimal places.

5. Click *OK* to proceed.

Now the attribute table shows an extra column with NULL values.

6. In order to calculate the correct values, click ℰ above the table to open the *Expression* dialogue. Fill in the dialogue as shown in figure 2.12. To avoid typos, the best practice is to double click on the field name in the middle of the dialogue screen and click the * button. Then type 0.1 so the equation is "TEMP_T(0.1C)" * 0.1. Click *OK* to proceed.

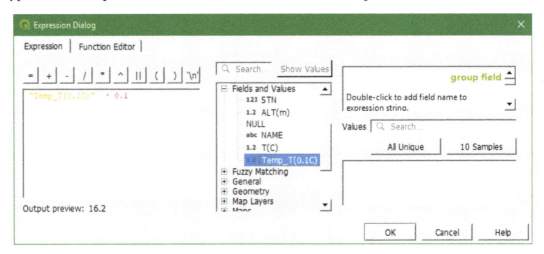

Figure 2.12: Expression Dialog of the Field Calculator

Make sure the *Attribute table* window looks like figure 2.13. Please note that T(C) should be selected as the column to which the calculated values will be assigned!

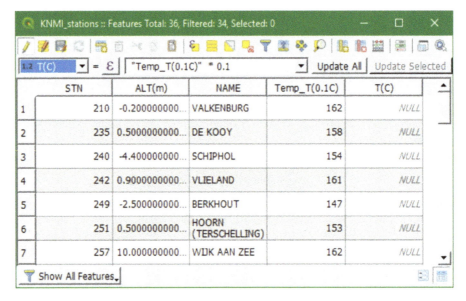

Figure 2.13: Field Calculator in the Attribute table

7. Click *Update All* to perform the calculation for all features.

Now check the result in the attribute table.

8. Click on the 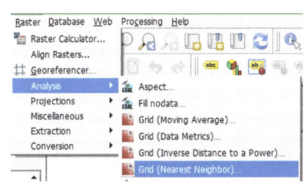 to toggle back to non-editing mode. Click *Save* to save the changes when asked and close the attribute table. If you made a mistake, don't save, but instead choose *Discard* to undo all changes since last save.

9. Now remove the table *KNMI Temperatures Table* from the layers list and check the attribute table of KNMI_stations.

- What columns do you see now?
- What can you conclude about the join function?

You could have saved the entire attribute table by saving *KNMI_ stations* to a new shapefile using the previously used Export | Save Features As function.

2.7 Interpolate Point Features to Raster

The final task is to interpolate the temperature values to a raster.

1. In the main menu choose Raster | Analysis | Grid (Nearest Neighbor). (figure 2.14)

Figure 2.14: Interpolate to raster using nearest neighbor

2. In the dialogue specify the output file: tday_NN.tif by using the browse window and specifying the .tif format.

3. Select T(C) as *Z value from field*. This is the field that we will interpolate to so-called *Thiessen polygons*.

4. Check the *Open output file after running algorithm* checkbox.

For the rest of the dialogue keep the defaults. The dialogue should now look like figure 2.15, on the following page.

Note that the dialogue generates a GDAL command. The tool is essentially a GUI for the gdal_grid command line tool (https://www.gdal.org/gdal_grid.html).

5. Click *Run* to proceed.

6. Click *Close* to close the dialogue.

7. Now repeat the interpolation using the Raster | Analysis | Grid (Inverse Distance to a Power) (IDW) algorithm. Call the result file tday_IDW.tif.

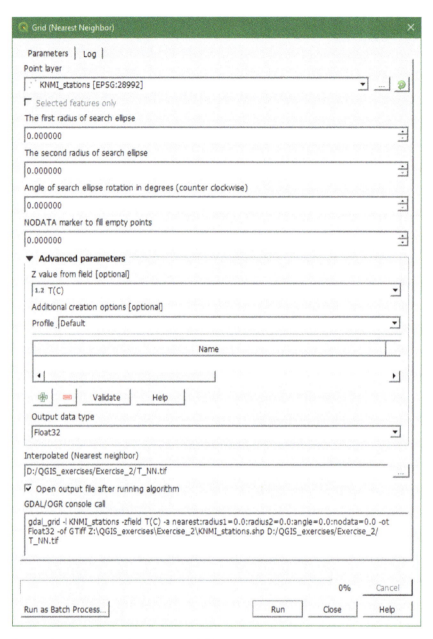

Figure 2.15: Interpolate to raster using nearest neighbor

2.8   Styling the Results

1. Drag the *KNMI_stations* layer so that it is positioned above both the *T_NN* and *T_IDW* raster layers in the *Layers Panel*.

To provide some context you will first add a basemap using the *QuickMapServices* plugin.

2. From the menu bar choose Web | QuickMapServices | OSM | OSM Standard.

3. Open the *Layer Styling panel* by clicking the ✎ button. Set the target layer to *KNMI_ stations*.

4. Select *Simple marker* component and set the *Fill color* to black and the *Size* to 2.8.

5. Switch to the *Labels* tab <abc> of the *Layer Styling Panel*. Switch from *No Labels* to *Single labels*.

6. Set the *Label with* option to the NAME field.

The names are all upper case. Next you will convert them to Title case.

7. Click the *Expression* ε button to open the *Expression Dialog* window. Use the Search window to find the title() function. Make the NAME field the string that is used with this function. Your expression will be title("NAME").

8. To add more information to the map you will now add the temperature to the label. Again click the *Expression* ε button. As you did in the previous exercise, use the *String Concatenation* ‖ operator and the *New line* 'n' operator to add the T(C) column to the label on a second line. Your labeling expression should now read title(NAME) || '\n' || "T(C)" || ' C'.

9. To center the label text switch to the label *Formatting* tab and set the *Alignment* to Center.

10. Set the *Font* to Calibri with a *Style* of Bold and a *Size* of 10.

11. Switch to the *Label buffer* tab abc and check the *Draw text buffer* option. Set the *Opacity* to 50%.

12. To give more separation between the labels and the feature icon, switch to the *Label placement* tab and set the *Distance* to 2.

If labels are being cut off at the map boundary, review the previous exercise to remember how to use the *Automated placement settings* to prevent it.

Now you will turn your attention to the two interpolated rasters.

13. Start by making the *T_NN* layer the target layer in the *Layer Styling Panel*.

14. Change the render from the default *Singleband gray* to *Singleband pseudocolor*.

15. For the *Color ramp* choose *Spectral*. Click *Classify*. By default this color ramp is setup to make the lower values red and the larger numbers blue. This is counter-intuitive. These values represent temperature so the higher values should be represented with warmer/red colors and the lower values with colder/blue colors. To accomplish this, right-click on the color ramp and choose *Invert Color Ramp* from the context menu. In order to see the basemap you will apply a **Blending mode**. In the *Layer rendering* section choose *Multiply* for the *Blending mode*.

Your map should resemble figure 2.16, on the next page.

16. Next turn off the *T_NN* layer and turn on the *T_IDW*. Use the *Layer Styling Panel* to style this using the *Singleband pseudocolor* renderer with the same inverted color ramp. This time switch the *Mode* to *Quantile*. Also apply a Multiply *Blending mode* to this raster.

Your map should now resemble figure 2.17, on page 49.

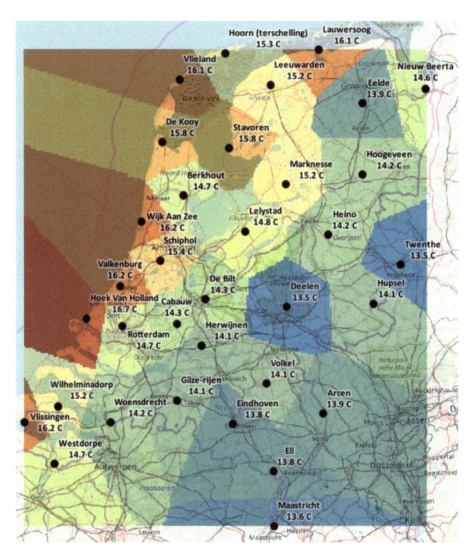

Figure 2.16: Stations and Nearest Neighbor Data Styled

- Which interpolation method is better? Why?

- Can you explain the temperature gradient in the map?

The following video explains the steps in this chapter: "Load spreadsheets in QGIS 3, join and interpolate". Videos can be found at the YouTube channel http://www.youtube.com/c/hansvanderkwast.

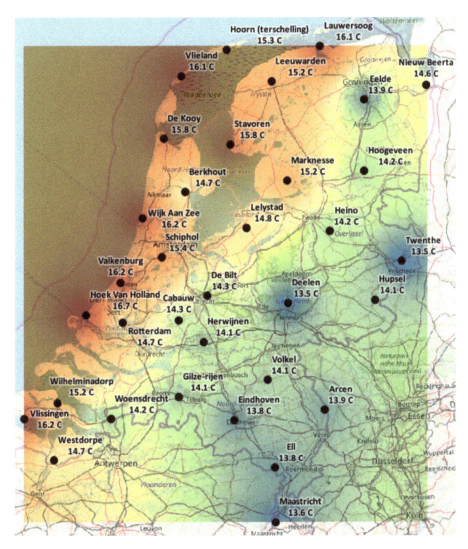

Figure 2.17: Stations and Inverse Distance Weighted Data Styled

3. Spatial Analysis with Map Algebra

3.1 Introduction

With *map algebra* we can do calculations with raster layers. This is useful for spatial analysis. For example, when we need to evaluate different criteria to find suitable or unsuitable locations we can use map algebra.

After this chapter you will be able to:

- apply *map algebra* for raster analysis

- distinguish *Boolean*, *discrete*, and *continuous* rasters

- make legends for Boolean, discrete, and continuous maps

- understand the use of *Nodata*

- use *logical operators*

- calculate *distances* from rasters

- *reclassify* rasters

- *convert* raster to vector points

- *sample* raster values with vector points

> For this chapter the following video is recommended as theoretical preparation: "GIS Raster Processing". Videos can be found at the YouTube channel http://www.youtube.com/c/hansvanderkwast.

Case Description

The municipality of the (imaginary) oasis Aïn Kju Dzjis has hired you to analyse which wells are unsuitable for its inhabitants based on the following conditions:

Condition 1: The wells should be within 150 meters of houses or roads.

Condition 2: No industry, mine, or landfill within 300 meters of the wells.

Condition 3: The wells should be less than 40 meters deep.

You will use map algebra to perform the required analysis.

Figure 3.1, on the next page shows the workflow for this chapter.

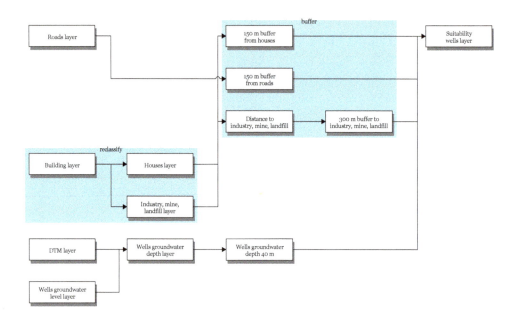

Figure 3.1: Workflow of this chapter

3.2   Preparation

Checking the Metadata

For this task you are provided with the following raster layers: `buildg.tif`, `roads.tif`, `dtm.tif`, and `gwlevel.tif`. This data can be downloaded from `http://locatepress.com/qgis_hydrological`.

First we are going to add the exercise folder to the *Favorites* in the *Browser Panel*.

1. In the *Browser* panel right-click on *Favorites* and choose `Add a directory` (figure 3.2).

Figure 3.2: Add Favorites in the Browser panel

2. Choose the folder where you have stored the layers.

3. Click the + to collapse the contents of the folder.

4. Preview the maps and metadata of these raster layers by right-clicking on the layer and choosing *Properties* (figure 3.3, on the next page).

The *Layer Properties* window will open, showing the metadata of the layer (figure 3.4, on the facing page).

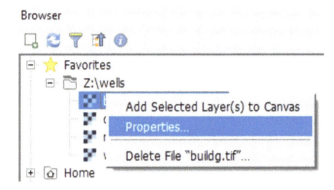

Figure 3.3: Layer properties menu

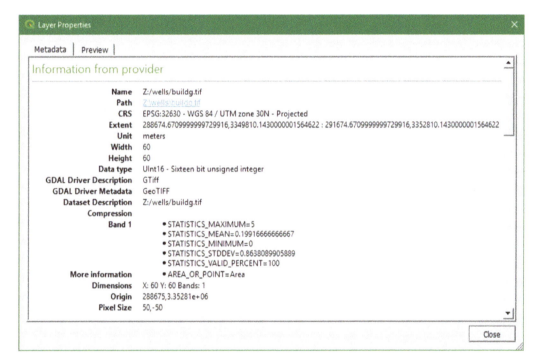

Figure 3.4: Layer properties window

- Check the metadata for all layers in this project. Note the file type, number of cells, projection, cell size, minimum value, and maximum value.

5. Select `buildg.tif`, `roads.tif`, `dtm.tif` and `gwlevel.tif` (keep the `Ctrl` button pressed and select the layers) and drag them to the *Map Canvas*.

6. Go to the *Layers* panel.

Using the Processing Toolbox

The *Processing Toolbox* in QGIS provides a lot of tools for processing GIS data. Besides QGIS tools, it also has tools from GDAL, GRASS, and SAGA that are very useful.

1. First activate the *Processing Toolbox* by choosing `Processing | Toolbox` from the main menu

(figure 3.5).

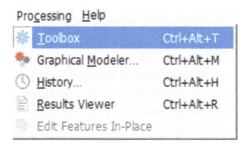

Figure 3.5: Open the Processing toolbox from the main menu

Next, we are going to change a default setting of QGIS. By default, the *Processing Toolbox* doesn't use the file name of the output as a layer name, which can be confusing.

2. In the main menu choose `Settings | Options| Processing`.

3. Collapse the `General` menu by clicking on the plus sign. Then check the box at *Use filename as layer name* (figure 3.6).

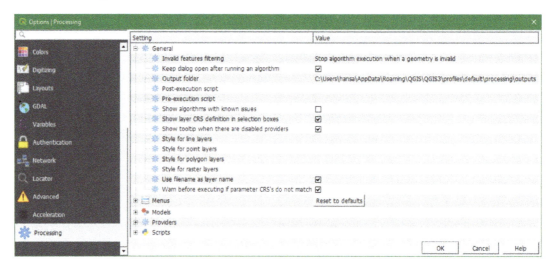

Figure 3.6: Activate use filename as layer name for processing tools

4. Click *OK* to close the dialogue.

3.3   Condition 1: Wells within 150 Meters of Houses or Roads

Let's first look at the houses. The houses are a class in the `buildg` layer. Rasters normally only store values, so we need to assign a proper legend to raster layers.

1. In the *Layers* panel click on the 🖌 to open the *Layer styling* panel.

2. Choose `Paletted/Unique values` as render type. Keep *Random* as *Color ramp* type and click *Classify*.

3. Use the screenshot of figure 3.7, on the next page as a reference to add the labels to the class numbers by double-clicking on each label name. Double-click on the color to make it more

intuitive as shown in the example.

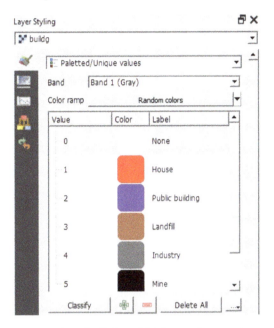

Figure 3.7: Styling a discrete raster using the Paletted/Unique values renderer

- Is buildg a Boolean, discrete, or continuous raster layer?

We have now learned that for discrete rasters we use the Paletted/Unique values render type.

Create a Boolean Layer with True for Houses and False for Other Buildings

If we want to create a Boolean layer with True (1) for houses and False (0) for the other classes in the buildg layer we can use the *Raster Calculator*.

1. In the main menu go to Raster | Raster Calculator.

2. In the *Raster Calculator* dialogue double-click on buildg@1, click the = button and type 1 (figure 3.8, on the following page).

Now the equation reads buildg@1 = 1, which means: if the buildg@1 layer equals 1 (which is the houses class), then the output layer has True (value 1), else it has False (value 0). @1 means band 1. In our case we only use single band raster layers (multiple bands are used in other applications such as remote sensing).

3. Call the output layer houses.tif and click *OK* to perform the calculation.

Following good practice, we are going to style this Boolean layer.

4. For Boolean layers we also use the Paletted/Unique values renderer. Repeat steps 1 and 2 of this subsection (3.3) and choose intuitive colors for True and False.

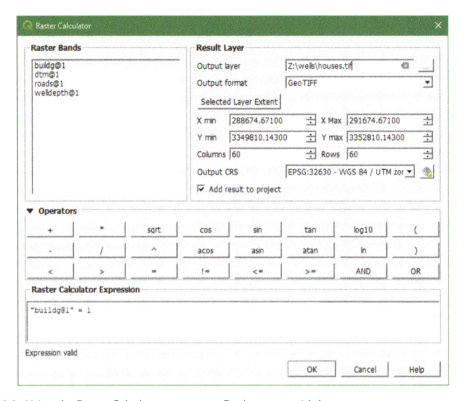

Figure 3.8: Using the Raster Calculator to create a Boolean map with houses

Create Zones of 150 Meters Around the Houses

Now that we have a layer with only houses, we can proceed with the calculation of the zones of 150 m around them. We're going to create a Boolean map which is True (1) within a zone of 150 m around houses and False (0) further than 150 m from houses.

1. In the main menu go to Raster | Analysis | Proximity (Raster Distance) (figure 3.9).

Figure 3.9: Proximity menu

2. In the *Proximity (Raster Distance)* dialogue make sure the houses layer is selected as *Input layer*. Set the *Distance units* to Georeferenced coordinates. Type for the *Maximum distance to be generated* 150 meters and type 1 for *Value to be applied to all pixels that are within the -maxdist of target pixels*. Set the *Output data type* to Byte (because we use only 0 and 1) and call the output *Proximity map* houses150m.tif. Leave the other settings at default (figure 3.10).

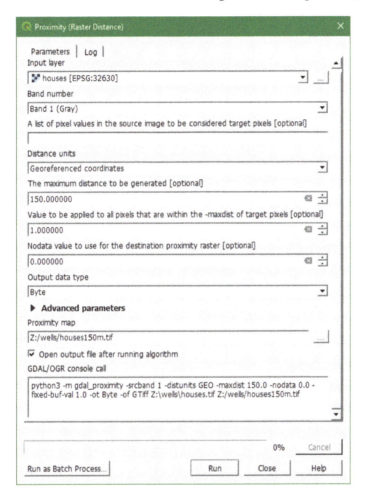

Figure 3.10: Proximity dialogue

3. Click *Run*, then *Close* when the calculation is done.

4. Style the Boolean map. Make the True pixels green and the False pixels red (figure 3.11, on the next page).

Create Zones of 150 Meters Around the Roads

In a similar way we can now calculate the 150 m buffer around the roads.

1. Style the roads layer.

2. Repeat the steps used to calculate the zones of 150 m around the houses, but use the roads layer as the *Input layer* and name the output roads150m.tif.

3. Style the Boolean map. Make the True pixels green and the False pixels red.

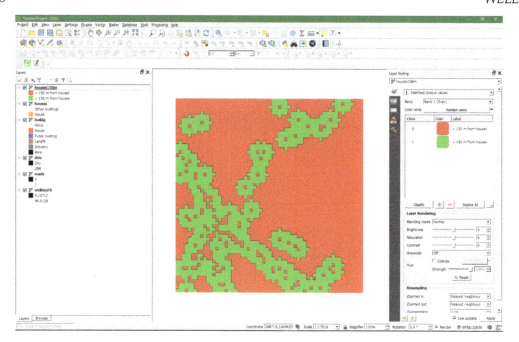

Figure 3.11: Styling of Boolean map with zone of 150 m from houses

3.4 Condition 2: No Industry, Mine, or Landfill within 300 Meters from Wells

For the second condition we first need to reclassify the `buildg` layer in such a way that the result is a Boolean map with True (1) for industry, mine, and landfill and False (0) for the other classes. Then we need to calculate the distance to the True pixels and finally calculate the pixels that are further than 300 m from industry, mine, and landfill.

Create a Boolean Raster with True for Industry, Mine, and Landfill, and False for Other Buildings

1. Choose in the *Processing Toolbox* `Raster analysis | Reclassify by table` (figure 3.12).

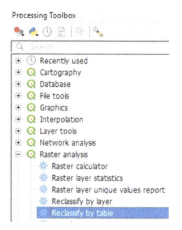

Figure 3.12: Reclassify by table processing tool

The *Reclassify table* dialogue appears. We are going to reclassify the `buildg` raster using a *lookup table*.

2. Fill in the dialogue exactly as in figure 3.13.

Here you can identify a nodata value for the output layer for values that are excluded from the lookup table. The *Range boundaries* define if values are included or excluded from the ranges in a row of the lookup table. Here we don't use ranges, but reclassify each value. Therefore we choose `min <= value <= max`. For the *Output data type* we use `Int16` because the output values are whole numbers.

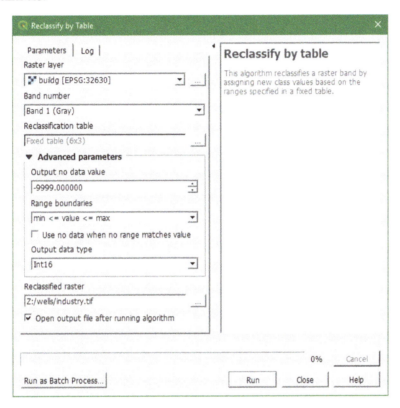

Figure 3.13: Reclassify by table dialogue

3. Then go to *Lookup table* and click ... .

4. Fill in the lookup table as shown in figure 3.14, on the following page.

5. Click *OK* and *Run*. *Close* the dialogue when the processing is finished.

6. Check the result: 1 for mines, industry, and landfills, 0 for the other classes. Use the Identify tool 🔍 and click on the map. On the lower right panel you can find the identify results. It displays the value of the pixel of the selected layer in the Layers panel. You might have to resize the columns to see the pixel values.

- Is `industry` a Boolean, discrete or continuous raster?

7. Style the `industry` layer.

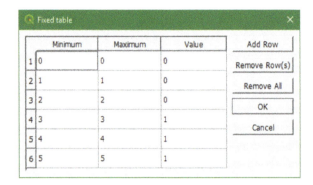

Figure 3.14: Lookup table for reclassification of raster values

Create Zones of 300 Meters Around Industry, Mine, and Landfill

With the Boolean map with True for industry, mine, and landfill we can calculate the distance to these pixels. Because the *Proximity (Raster Distance)* tool does not allow us to assign values for pixels larger than a threshold, we have to calculate all distances and then use the *Raster Calculator* to calculate a Boolean map with True for pixels further than 300 meters from industry, mine, and landfill.

1. Open the *Proximity (Raster Distance)* tool.

2. Make sure that Industry is the *Input layer*. Keep all other settings as default. Name the *Output proximity map* inddist.tif (figure 3.15, on the next page).

3. Check the result.

   - Is the inddist layer a Boolean, discrete, or continuous raster?

4. Make a legend that is appropriate for this raster type using *Singleband pseudocolor* as render type. Use intuitive colors (e.g. a ramp from red to green).

5. Use the *Raster Calculator* to calculate inddist@1 >= 300. Call the output layer ind300m.tif

   - Is the ind300m layer a Boolean, discrete, or continuous raster?

6. Style the ind300m layer (figure 3.16, on the facing page).

3.5   Condition 3: Wells Less than 40 Meters Deep

For the last condition we need to identify the wells that are less than 40 m deep. The gwlevel layer gives the absolute elevation of the groundwater level in the well in meters above sea level. In order to calculate the depth to the groundwater, we need to subtract this from the surface elevation given in the digital terrain model (DTM).

1. Open the *Raster Calculator*.

2. Subtract the absolute well depth from the DTM using this calculation: dtm@1 - gwlevel@1. Call the output layer welldepth.tif.

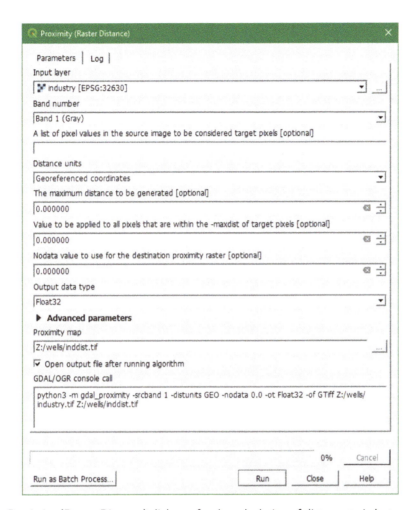

Figure 3.15: Proximity (Raster Distance) dialogue for the calculation of distance to industry, mines, and landfill

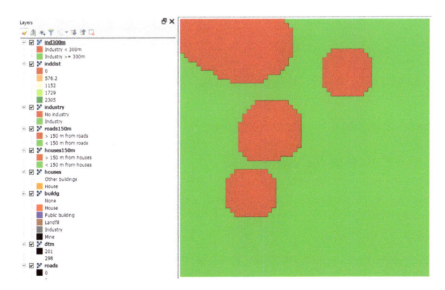

Figure 3.16: Result of the second condition (>= 300 m from industry, mines or landfills)

- Is the welldepth layer Boolean, discrete, or continuous?

3. Style the welldepth layer.

4. Next, calculate in the *Raster Calculator* a Boolean map with wells less than 40 m deep. Call the output layer notdeep.tif.

5. Style the notdeep layer (figure 3.17).

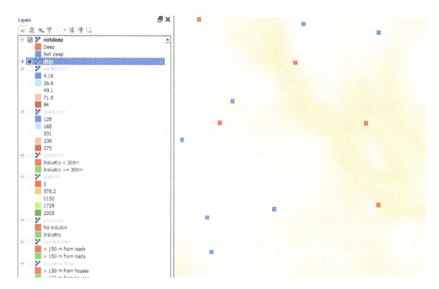

Figure 3.17: Result of the third condition (well depth < 40 m)

3.6   Combine the Three Conditions and Report the Results

Combine the Three Conditions

After calculating the Boolean maps for the three conditions we need to combine them to come to the final result.

1. Use the *Raster Calculator* to combine the three conditions. Because all Boolean results for the conditions need to be true we have to use the AND operator (figure 3.18, on the facing page). Call the output layer accessiblewells.tif.

2. Check the resulting raster layer and style the layer.

Convert Raster Cells to Point Vectors

To present the end result we can style the accessiblewells layer. However, it is nicer to present the wells as point features on the map. Therefore we need to convert the well pixels to point vectors.

1. In the *Processing Toolbox* go to Vector creation | Raster pixels to points (figure 3.19, on the next page).

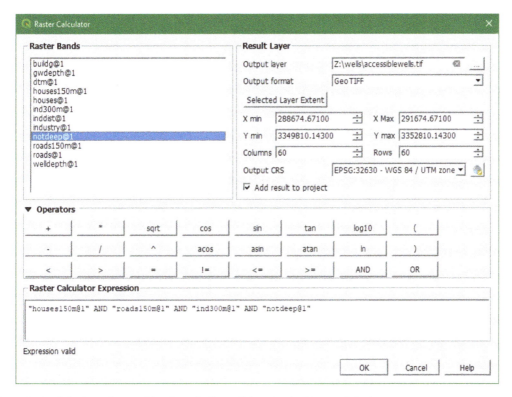

Figure 3.18: Calculate the combination of all conditions in the Raster Calculator

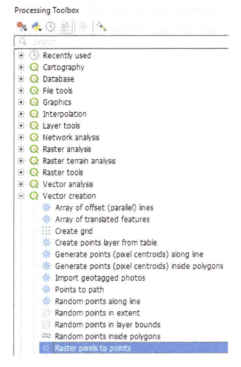

Figure 3.19: The raster pixels to points processing tool

2. In the dialogue choose the accessiblewells layer as the input *Raster layer*. For *Field name*, enter Accessibility. The raster values will be saved under this field in the attribute table. Choose wells.shp as the output *Vector points* layer (figure 3.20).

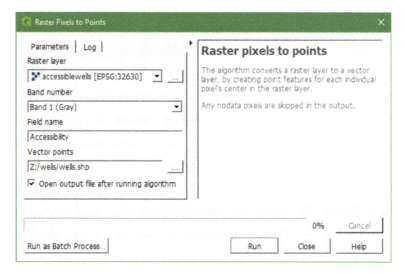

Figure 3.20: The raster pixels to points processing tool dialogue

3. Click *Run*, then *Close* when the conversion is finished.

Sample Raster Values

The point vector layer wells now only contains the field Accessibility. It is however, more informative to also include other data in the attribute table. With the *Point sampling tool* plugin we can sample the raster layers in this project and add that information to the point attribute table.

1. Install the *Point sampling tool* plugin.

2. In the *Layers* panel only check the boxes before the layers you want to sample from and uncheck the others. Choose the following layers: dtm, welldepth, gwdepth, notdeep, ind300m, roads150m, and houses150m.

3. Click the *Point sampling tool* button .

4. In the *General* tab of the *Point Sampling Tool* dialogue choose wells.shp as the *layer containing sampling points*, select all *Layers with fields/bands to get values from* and save the *Output point vector layer* to wells_final.shp (figure 3.21, on the next page).

5. Click the *Fields* tab.

6. Edit the *name* of the attribute that will be given to the output field if needed (figure 3.22, on the facing page).

7. Click *OK* and *Close*.

8. Open the attribute table of wells_final and check the result (figure 3.23, on page 66).

Figure 3.21: The General tab of the Point Sampling Tool dialogue

Figure 3.22: The Fields tab of the Point Sampling Tool dialogue

Style the Analysis Results

1. The first step in showing the results will be to style the wells_final data by whether the wells are accessible or not. In the *Layers* panel click on the ✎ to open the *Layer styling* panel. Set wells_final as the target layer.

2. Change from a *Single symbol* to a *Categorized* renderer. Then set the *Column* to Accessible. Click *Classify*.

3. When using this renderer QGIS will create a category to capture any NULL values. Here there are no wells with NULL values in the Accessible column. Select the entry where the

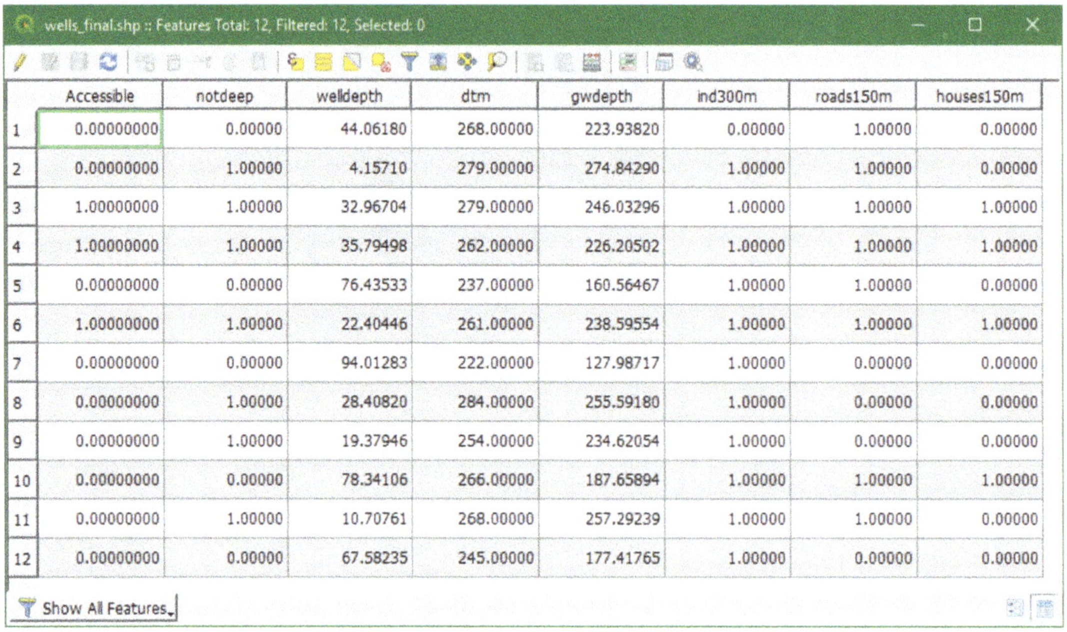

wells_final.shp :: Features Total: 12, Filtered: 12, Selected: 0

	Accessible	notdeep	welldepth	dtm	gwdepth	ind300m	roads150m	houses150m
1	0.00000000	0.00000	44.06180	268.00000	223.93820	0.00000	1.00000	0.00000
2	0.00000000	1.00000	4.15710	279.00000	274.84290	1.00000	1.00000	0.00000
3	1.00000000	1.00000	32.96704	279.00000	246.03296	1.00000	1.00000	1.00000
4	1.00000000	1.00000	35.79498	262.00000	226.20502	1.00000	1.00000	1.00000
5	0.00000000	0.00000	76.43533	237.00000	160.56467	1.00000	1.00000	0.00000
6	1.00000000	1.00000	22.40446	261.00000	238.59554	1.00000	1.00000	1.00000
7	0.00000000	0.00000	94.01283	222.00000	127.98717	1.00000	0.00000	0.00000
8	0.00000000	1.00000	28.40820	284.00000	255.59180	1.00000	0.00000	0.00000
9	0.00000000	1.00000	19.37946	254.00000	234.62054	1.00000	0.00000	0.00000
10	0.00000000	0.00000	78.34106	266.00000	187.65894	1.00000	1.00000	1.00000
11	0.00000000	1.00000	10.70761	268.00000	257.29239	1.00000	1.00000	0.00000
12	0.00000000	0.00000	67.58235	245.00000	177.41765	1.00000	0.00000	0.00000

Show All Features

Figure 3.23: The final attribute table for the wells

Value reads *all other values* and click the *Delete* button to remove it. Give the wells with a value of 1 a green symbol with a size of 4. Give the wells with a value of zero a red color with a size of 2.

4. In the *Legend* column rename the entry with a *Value* of 1 to *Accessible* and the entry with a *Value* of zero to *Inaccessible* (figure 3.24).

Figure 3.24: Styling the Final Wells

5. Scroll down in the *Layer styling* panel and find the *Layer Rendering* section. Expand it and select *Draw effects*. The *Customize effects* icon becomes active. Click on it to open the *Effects properties* panel. Here you can add Inner and Outer Glows, Drop Shadows, and other effects. Click the box next to *Drop Shadow*. Select the *Drop Shadow* effect so that you are seeing the parameters of the *Drop Shadow*. Reduce the *Offset* distance to 1.0 (figure 3.25, on the next page). Click the *Go back* button to return to the main styling panel.

6. Switch to the *Labels* tab of the *Layer styling* panel.

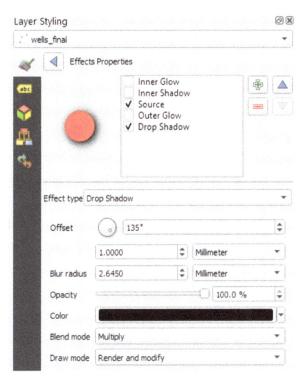

Figure 3.25: Setting Drop Shadow Parameters on the Effects Panel

7. Change the setting from *No labels* to *Single labels*. Set the *Label with* column to welldepth.

8. You will use a labeling expression very similar to that used at the end of the "Preparing Data from Hardcopy Maps" chapter, on page 29. You will set the expression up so the labels read something like: *Well Depth: 76.4*.

9. Click the *Expression* [ε] button to open the *Expression Dialog* window.

10. To begin, add the string *Well Depth* in front of the well depth value by adding 'Well Depth: ' before the existing expression.

11. To accomplish the second component you will use the format_number function. Use the search box to find the format_number function. Insert it right before the welldepth field. Remember, this function requires a number and a number of decimal places. The number will be the welldepth field and the number of places 2.

12. Use what you learned about the *String Concatenation* [||] operator and the the *New line* ['\n'] operator to nicely format the label.

13. Your expression should resemble 'Well Depth:'|| '\n' || format_number(welldepth,2).

14. Set the *Font* to a sans serif font with a *Size* of 9.

15. Switch to the *Placement* tab and set the *Placement* to *Offset from point*. Select the lower right *Quadrant* placement with an *Offset X Y* setting of 1.5 each (figure 3.26, on the next page).

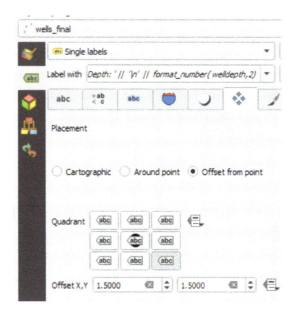

Figure 3.26: Label Placement Settings

16. Finally click the *Automated placement settings* button to open the *Automated Placement Engine* window. Uncheck the box for *Allow truncated labels on edges of map* option. This will prevent labels from being cut off.

17. To complete the styling you will work with the dtm layer. Make this layer the target layer in the *Layer styling* panel. Click the drop-down for the *Color ramp* and choose *Create new color ramp*.

18. The *Color ramp type* window opens. Choose *Catalog:cpt-city* as the type. Click *OK*.

19. The *Cpt-city Color Ramp* window will open. Select the *Topography* category.

20. Choose *sd-a.* and click *OK*.

> Did you know that you can save this color ramp to your style library? To do this access the color ramp context menu and select *Save color ramp*. The *Save New Color Ramp* window will open allowing you to give it a *Name* and provide *Tag(s)*.

21. Scroll down to the *Layer Rendering* section and set the *Blending mode* back to *Normal*.

22. Finally right-click on the dtm layer and choose *Duplicate* from the context menu.

23. Drag the dtm copy layer above the dtm layer and turn it on.

24. Make this duplicated dtm layer the target layer in the *Layer styling* panel. Change the renderer to *Hillshade*. Scroll down to the *Layer Rendering* section and set the *Blending mode* back to *Multiply* (figure 3.27, on the facing page).

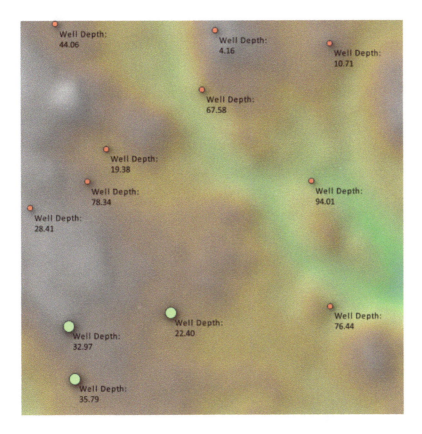

Figure 3.27: Final Analysis Results Styled

4. Stream and Catchment Delineation

4.1 Introduction

One of the most important uses of GIS in hydrology is the delineation of catchments and streams. This chapter presents a generic workflow for stream and catchment delineation for areas where only open data is available. The Shuttle Radar Topography Mission (SRTM) 1 Arc Second DEM will be used. The workflow will be applied to the Rur catchment in Germany.

After this chapter you will be able to:

- find and *download SRTM DEM tiles* from the EarthExplorer website

- *mosaic raster layers* into a virtual raster

- *reproject* rasters

- *create subsets* of rasters

- *fill sinks* in a DEM

- calculate and style *flow direction* raster layers

- calculate *Strahler orders*

- *delineate streams*

- *delineate catchments*

- style layers for visualising catchments

Generic Workflow for Catchment Delineation

In order to delineate a catchment from a DEM in GIS we need to follow these steps:

1. Download the DEM tiles of your study area. Make sure that the tiles cover at least the study area and that the catchment you want to derive is covered completely. Better to take it a bit larger to avoid boundary effects.

2. If your study area is covered by multiple DEM tiles, you need to mosaic (merge) the tiles to create a single raster DEM layer.

3. The DEM tiles might be in a different coordinate system then desired. In that case you have to reproject the DEM layer to the projection you want to use in the study area.

4. In the case that the merged tiles are much larger than your study area, you can subset (clip) it to a smaller area to reduce calculation time.

5. Interpolate voids when pixels with no data exist in the DEM.

6. Make a hydrologically correct DEM by filling sinks and removing spikes.

7. Calculate the flow direction for each cell.

8. Derive the drainage network.

9. Calculate the catchment for the outflow point of the catchment.

When a map with the stream network already exists, the procedure can be improved by "burning" the river network into the DEM. In that way the DEM is always lower at rivers and runoff will follow the actual river network. This is beyond the scope of this chapter.

Figure 4.1 shows the workflow.

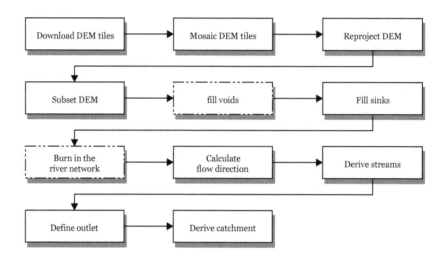

Figure 4.1: Workflow for stream and catchment delineation

Data for this chapter can be downloaded from http://locatepress.com/qgis_hydrological. The next section covers how to download the SRTM tiles from USGS EarthExplorer. These can also be downloaded from the course data repository.

> For this chapter the following video is recommended as theoretical preparation: "Stream and Catchment Delineation with GIS". Videos can be found at the YouTube channel http://www.youtube.com/c/hansvanderkwast.

4.2 Download DEM Tiles

For the Rur study area, we will download the tiles from the SRTM 1 Arc-Second global data set. Since the end of 2014 a 1-arc second global digital elevation model (approximately 30 meters at the equator) has been released as open data. Most parts of the world have been covered by this dataset ranging from 54 degrees south to 60 degrees north latitude. A description of the SRTM data products can be found here: http://loc8.cc/srtm1arc.

The following steps explain how to download the SRTM DEM tiles for the study area.

1. Open a browser and browse to `http://earthexplorer.usgs.gov`

In order to download the SRTM DEM tiles you need to create an account.

2. Click *Register* and follow the procedure. After registration, log in to the USGS EarthExplorer portal.

Now we need to indicate the search area. In EarthExplorer this can be done in different ways: (1) Search for a location name, (2) Zoom in on the map and use the coordinates of the zoom area, (3) Upload a KML or zipped shapefile. For this case we prepared a shapefile with a bounding box.

3. Choose *Shapefile* and upload the `boundingbox.zip` file (see figures 4.2 and 4.3).

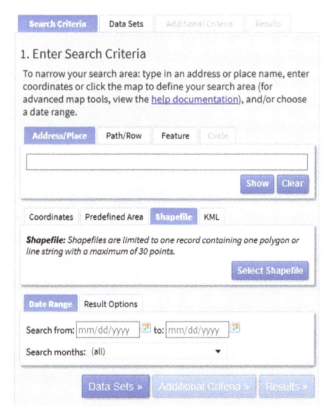

Figure 4.2: Use a Shapefile to indicate the search area

4. Choose *Data Sets* to go to the next screen where we can select the data set.

5. Click on the plus signs to select `Digital Elevation | SRTM | SRTM 1-Arc Second Global`. Make sure to tick the box (figure 4.4, on page 75).

6. Click *Results* to go to the next screen.

Now you see the search results: four tiles. Before downloading you can check the tiles. With the foot icon, for example, you can see the footprint of the tile (figure 4.5, on page 76).

7. Download each tile by clicking the *Download Options* button ⬇ .

8. In the *Download Options* dialogue download the DEM in GeoTIFF Format to the exercise

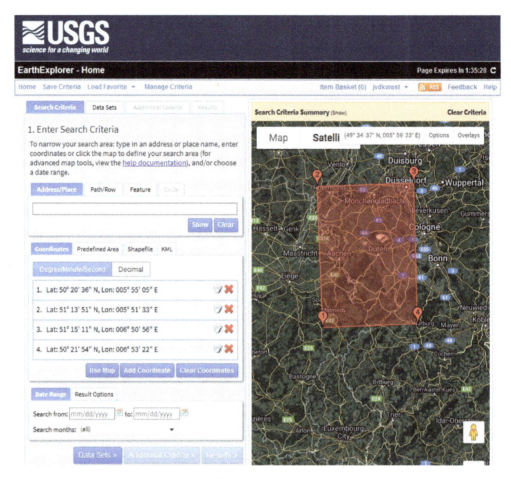

Figure 4.3: Search criteria with loaded Shapefile

folder on your hard disk (figure 4.6, on page 76).

The tiles can now be added to your QGIS project.

9. Start QGIS Desktop. We'll start with a new project. In the browser panel add the folder with the data for this chapter to the *Favorites*.

10. Right-click on one of the tiles and choose `Properties` to review the metadata (figure 4.7, on page 77). What is the projection? What are the map units?

11. Select and drag the four tiles to the Map Canvas (figure 4.8, on page 77).

4.3 Mosaic DEM Tiles

Before we proceed, we have to merge the four DEM tiles, which in GIS terminology is called mosaic. Here are two ways to mosaic the tiles:

- Merge the tiles into one physical file (e.g. GeoTIFF)

- Merge the tiles into a virtual file (`.vrt`)

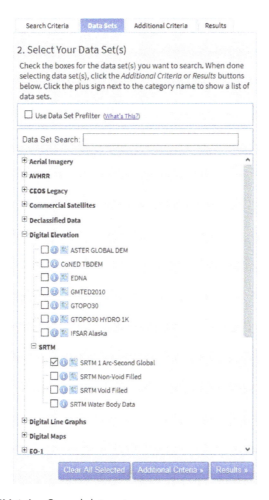

Figure 4.4: Select the SRTM 1 Arc Second data set

The first option is slower. If we have many tiles, we prefer to create a file that virtually merges all the tiles. That will be done with the following steps:

1. In the main menu choose `Raster | Miscellaneous | Build Virtual Raster` (figure 4.9, on page 77)

2. In the *Build Virtual Raster* dialogue you can choose each file individually or merge all files in a directory. We can also merge the files that are visible in the Map Canvas. We'll use the last option:

- At Input layers click `...` .

- Use the *Select all* button to select the four tiles and click *OK* (figure 4.10, on page 78).

- Browse to the location where you want to save the output file and give it the name dem_ mosaic.vrt

- By default, *Resolution* is set to average. In our case the files all have the same resolution (1 Arc Second).

- Uncheck the box before *Place each input file into a separate band*. This needs to be checked only if you want to create a mapstack, i.e. with remote sensing bands.

Figure 4.5: Search result

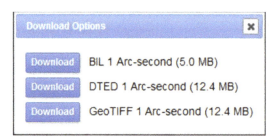

Figure 4.6: Download options dialogue

- Keep the *Resampling algorithm* at the default: nearest. This also has no impact, because resampling is not needed.

The dialogue should now look like that shown in figure 4.11, on page 78.

3. Click *Run* to run the algorithm. Click *Close* to get back to the main screen where you can see the merged DEM.

You'll notice that in the Map Canvas, the borders of the tiles are not visible anymore in the

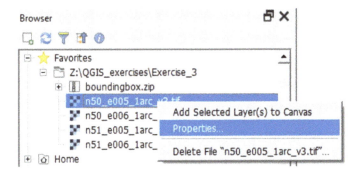

Figure 4.7: Show properties of a layer

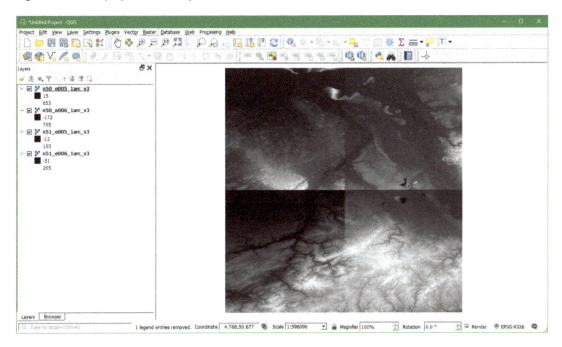

Figure 4.8: Four SRTM tiles

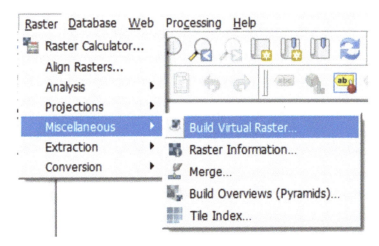

Figure 4.9: Build virtual raster

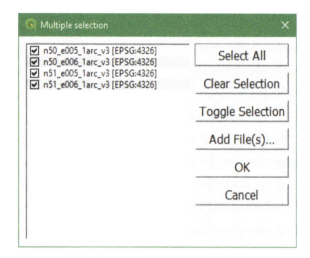

Figure 4.10: Multiple selection dialogue

Figure 4.11: Build virtual raster dialogue

merged DEM because QGIS stretches the greyscale using the minimum and maximum of the entire merged DEM. This is only for visualisation, the values in the tiles are the same as in the mosaic.

4. Now remove the individual tiles (not the dem_mosaic) from the layers list by selecting them while the Ctrl button is pressed. Then right-click one of the tile names and select Remove Layer. Click *OK* to confirm. This will remove the tiles from the screen, but not from the hard disk (figure 4.12, on the facing page).

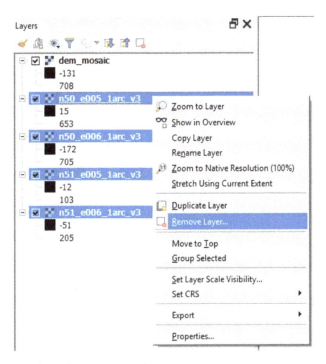

Figure 4.12: Remove layers from the Layers Panel

4.4 Reproject DEM

The DEM is in its original Lat/Lon Geographic Coordinate System (GCS) with datum WGS 84 (EPSG: 4326). It is not recommended to use a GCS for DEM analysis, because the z units (e.g. meters) are different than the x and y units (degrees). We need to choose a projection for our project. If the project covers one country we can choose a national projection. In our case, however, the project covers multiple countries (Germany, the Netherlands, and Belgium). Therefore we will use a global projection: UTM Zone 32 North, with WGS-84 as datum.

We can find the EPSG codes at http://www.spatialreference.org

1. Use the website to search for UTM 32N wgs 84. You can leave QGIS running and open a browser.

Figure 4.13, on the next page shows the result.

We need to take a look at the EPSG codes. We will use EPSG: 32632 throughout this project—clicking on it will provide more details.

Now we are going to reproject the DEM from unprojected GCS (Lat/Lon WGS 84 - EPSG: 4326) to UTM Zone 32 North / WGS 84 (EPSG: 32632).

2. From the main menu choose Raster | Projections | Warp (Reproject) (figure 4.14, on the following page)

3. In the *Warp (reproject)* window, click to choose the *Target CRS*.

4. In the dialogue that opens type 32632 at *Filter* and select WGS EPSG: 32632 in the middle of the dialogue window and click *OK* (figure 4.15, on page 81).

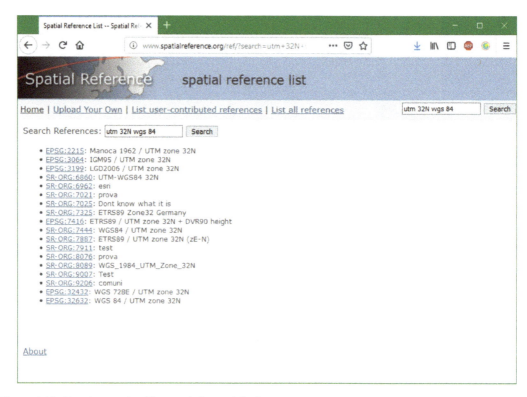

Figure 4.13: Results search of keywords in spatialreference.org

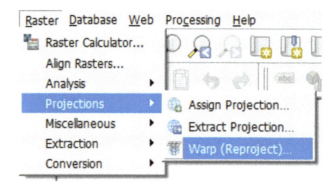

Figure 4.14: Reproject rasters using Warp

5. Now complete the dialogue:

- For the *Resampling* method we choose `Nearest Neighbor` to preserve the elevation values of the original files.

- Set the *No data value for output bands* to `-9999` because when the raster layer is reprojected there will be "no data" at the borders. In this way we define that "no data" has a value -9999 and will not be visualised and treated as "data".

- Set the *Output file resolution* to `30` m.

- Browse to your exercise folder and name the output file `dem_reprojected.tif`

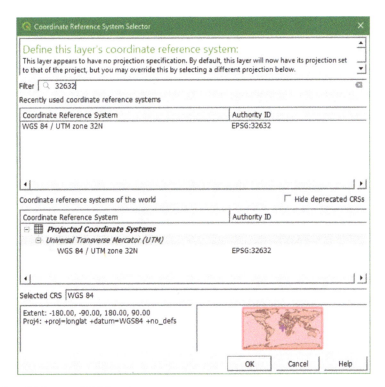

Figure 4.15: Select CRS using EPSG code

The dialog should now look like that shown in figure 4.16, on the following page. Note the `gdalwarp` command that will be executed in the background.

6. Click *Run* to run the algorithm. After running, click *Close* to close the window.

The re-projected DEM now appears in the list of Layers. The DEM may seem distorted due to the fact that the Map Canvas Projection is still in Lat/Lon, as indicated by the coordinates in the lower part of the Map Canvas and the EPSG code of the Map Canvas in the lower right corner. This is due to the on-the-fly projection, which causes a difference between the projection in the file and the one that is visualised.

7. To complete this operation, and properly display the new dataset, we need to change the on-the-fly projection of the Project. To do so click on the EPSG:4326 in the lower right of the screen (figure 4.17, on the next page).

8. In the new dialogue window select the `EPSG:32632` projection that is already in the list of *Recently used coordinate reference systems*, as shown in figure 4.18, on page 83, and click *OK*.

In the Map Canvas the DEM is now displayed in the correct position (with north pointing up), as shown in figure 4.19, on page 84. The coordinates of the new display are also in meters as we want.

9. You can now remove the layer `dem_mosaic`

This is a good time to save your project.

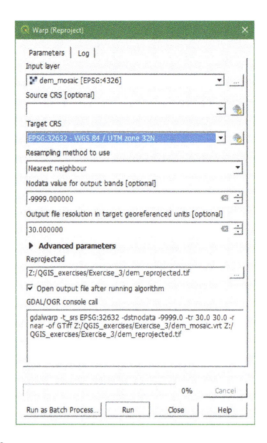

Figure 4.16: Warp dialogue

Figure 4.17: Map Canvas coordinates

4.5   Subset DEM

In order to reduce the calculation time of the algorithms, we will subset (or clip) the raster layer.

> Always make the boundary of your study area a little bit bigger to prevent boundary effects.

An easy way to select the boundary of your study area is to use OpenStreetMap. OpenStreetMap contains crowd sourced data (see http://www.openstreetmap.org for more info). If the *QuickMapServices* plugin is installed, you can add an OpenStreetMap to QGIS as follows:

1. In the main menu choose Web | QuickMapServices | OSM | OSM Standard.

2. Now the OpenStreetMap layer will be shown in the Map Canvas. Temporarily hide the DEM by unchecking the box or dragging the OSM layer to the top.

In order to know the approximate extent of the Rur catchment, the exercise data contains a zipped shapefile with the bounding box (the same that we used for downloading the DEM tiles).

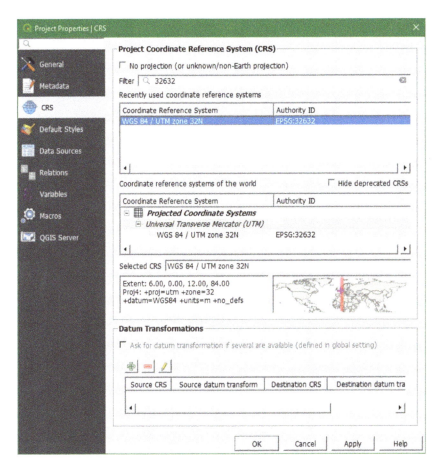

Figure 4.18: Change the on-the-fly reprojection

3. Unzip `boundingbox.zip` in your exercise folder.

4. Drag the `boundingbox.shp` layer to the Map Canvas (figure 4.20, on the following page).

5. Now, from the main menu select `Raster | Extraction | Clip Raster by Mask Layer` (figure 4.21, on page 85).

6. In the *Clip Raster by Mask Layer* dialogue, choose `dem_reprojected` for the *Input layer*. For *Mask Layer*, choose `boundingbox`. Keep the defaults for the other options. Choose `dem_subset` for the *Clipped (mask)* output and click *Run*. Click *Close* when done (figure 4.22, on page 85).

> Often you'll not have a bounding box shapefile. In that case you can choose `Raster | Extraction | Clip Raster by Extent`. Then you can drag a box in the Map Canvas and use that for clipping. While using that, make sure your on-the-fly reprojection is similar to the layer that you want to clip, because the Map Canvas coordinates are used by the algorithm.

7. Now you can remove `dem_reprojected` from the layers list as we have done before for other layers that are no longer needed.

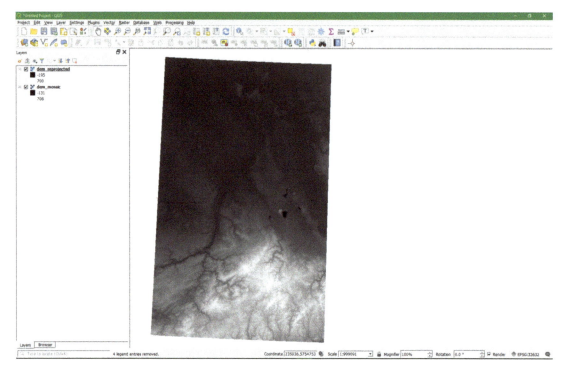

Figure 4.19: DEM in the right OTF reprojection

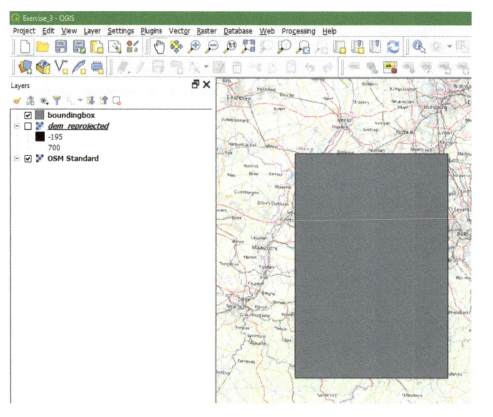

Figure 4.20: Bounding box polygon

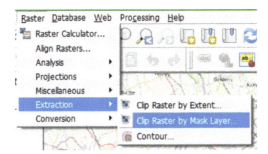

Figure 4.21: Clip raster by mask layer

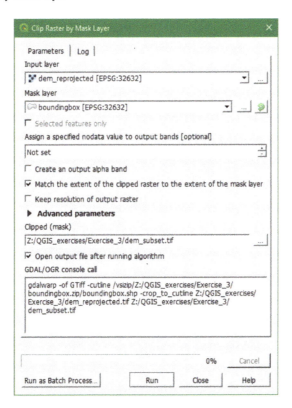

Figure 4.22: Clip raster by mask layer dialogue

4.6 Fill Sinks

Raw, unprocessed DEMs have artifacts such as depressions. Artifacts are a result of the DEM acquisition process and must be removed before a DEM can be used for hydrological analysis, like catchment and stream delineation or hydrological modelling. There are several algorithms for filling sinks. We will use the algorithm developed by Wang and Liu (2006), which is faster than others and works better with high resolution datasets. This algorithm, implemented in the SAGA tools, not only fills the depressions, but also preserves a downward slope along the flow path. This is achieved by preserving a minimum slope gradient between cells.

1. We will use the QGIS *Processing Toolbox*. If the *Processing Toolbox* isn't visible at the right of your screen, enable it by choosing Processing | Toolbox from the main menu (figure 4.23, on the next page).

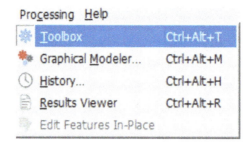

Figure 4.23: Open the Processing toolbox from the main menu

2. In the *Processing Toolbox* use the *Search* field to search for fill sinks (figure 4.24).

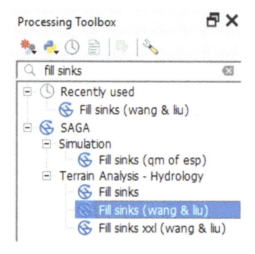

Figure 4.24: Search for fill sinks in the Processing toolbox

> You will notice there are more algorithms for filling sinks. Each has its own advantages and disadvantages in terms of calculation time, memory usage, and accuracy. For your own research you need to make a decision based on the documentation of the algorithms or by analysing the results of each.

3. Go to SAGA | Terrain Analysis - Hydrology | Fill sinks (wang & liu).

4. In the dialogue, keep the default *Minimum Slope [Degree]* of 0.001. Make sure to select dem_ subset as the input *DEM* and define dem_fill.sdat as the *Filled DEM*. We only need the *Filled DEM* at this point, so you can uncheck the boxes for opening the output of *Flow Directions* and *Watershed Basins* (see figure 4.25, on the next page).

5. Click *Run* and *Close* when done.

6. Now remove dem_subset from the layers list, as it is no longer needed.

> The default legends produced by Desktop GIS software are often not intuitive, because the software does not know (1) what information is presented, and (2) the raster data type (e.g. Boolean, discrete or continuous data). To help interpret the results it is good practice to style your layers intuitively.

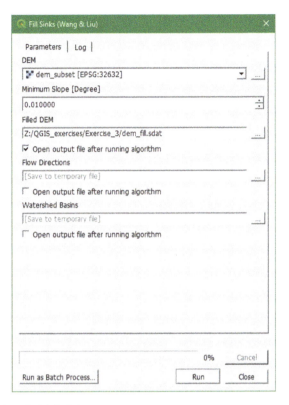

Figure 4.25: Fill sinks (Wang & Liu) dialogue

The DEM is a continuous raster. Continuous rasters represent gradients and can therefore contain real numbers (or floating point). Continuous rasters are styled in QGIS using ramps in the *Singleband pseudocolor* dialogue.

7. Select the dem_filled layer and click ✎ to open the *Layer Styling* panel (or press F7).

8. In the *Layer Styling* panel choose *Singleband pseudocolor* from the drop-down list (figure 4.26, on the following page).

The legend is automatically updated with the default red to blue color ramp. This obviously gives the wrong impression that there are lakes in the south, because we associate blue with lower areas with water. In the next steps we will apply more intuitive colors for DEMs.

9. Right-click on the color ramp and choose Create New Color Ramp (figure 4.27, on the next page).

10. In the popup *Color ramp type* dialogue choose *Catalog: cpt-city* from the drop-down list (figure 4.28, on the following page).

The cpt-city catalogue has a lot of useful preset color ramps.

11. Choose Topography | Elevation. Note that *cd-a* and *sd-a* are also nice choices (figure 4.29, on page 89).

This gives us more intuitive colors in the DEM where we can clearly distinguish higher and lower areas. Now we will further improve the visualisation.

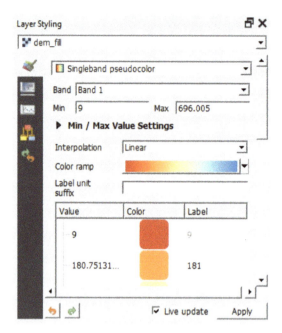

Figure 4.26: Layer styling panel for Singleband pseudocolor

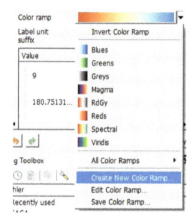

Figure 4.27: Create new colour ramp

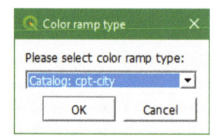

Figure 4.28: Color ramp type dialogue

12. Right-click on the dem_fill layer and select Duplicate Layer (figure 4.30, on the facing page).

This creates a copy of the dem_fill layer called dem_fill copy.

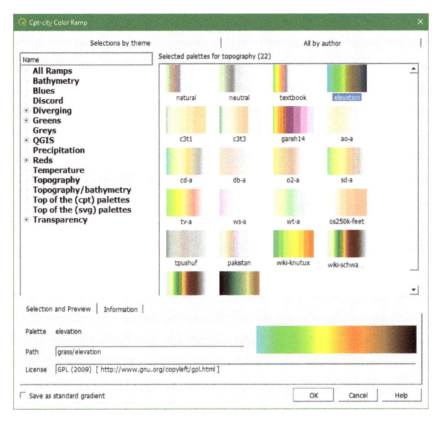

Figure 4.29: Cpt-city dialogue

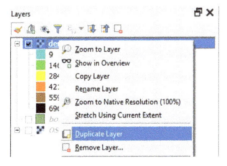

Figure 4.30: Duplicate layer

13. Uncheck the dem_fill layer and rename the dem_fill copy layer to hillshade.

14. In the *Layer Styling* panel, which should still be open, make sure that the hillshade layer is now selected. In the drop-down list change *Singleband pseudocolor* to *Hillshade*.

Now the hillshade layer is visualised with a shading. Which direction is the illumination coming from? Is this possible in reality?

Hillshade gives the best results with an artificial illumination in the northwest, which in reality can not exist in the Northern Hemisphere. If you move the dial in the *Layer Styling* panel to the southwest, you will see an inverted relief. Also note that there is a *Resampling* section. The default resampling method for both Zoomed in and Zoomed out is Nearest Neighbor. This method is fine for categorical data however, elevation is considered continuous data. You should therefore choose a *Zoomed in* resampling method of *Bilinear* and a *Zoomed out* resampling method of *Average*.

Next, we're going to blend the DEM with the hillshade layer.

15. Switch on the dem_fill layer by checking the box.

16. In the *Layer Styling* panel make sure the dem_fill layer is selected. In the *Layer Rendering* block of the panel, change the *Blending mode* to *Multiply* (figure 4.31).

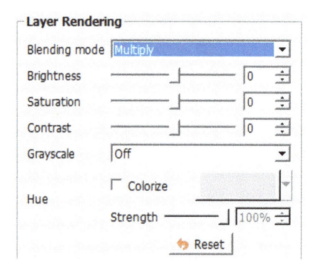

Figure 4.31: Change the blending mode

As you can see, blending gives a much nicer effect than transparency. With transparency the colors will fade. Now we can clearly see the elevation differences: the gradient from south to north and the valleys where we expect the streams. (see figure 4.32, on the facing page).

Blending modes allow for more elegant rendering between GIS layers. They can be much more powerful than simply adjusting layer Opacity. Blending modes allow for effects where the full intensity of an underlying layer is still visible through the layer above. There are 13 blending modes available. See https://docs.qgis.org/3.4/en/docs/user_manual/introduction/general_tools.html#blend-modes.

4.7 Delineate Streams

Calculate Strahler Orders

Before we can derive the streams from the DEM, we need to determine what we consider streams. For this purpose we use the Strahler order. The higher the order, the bigger the stream (see figure 4.33, on the next page).

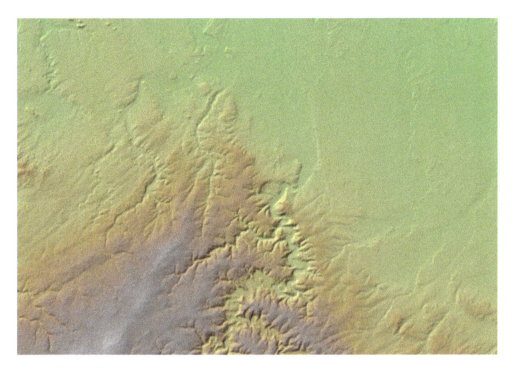

Figure 4.32: Color Hillshade

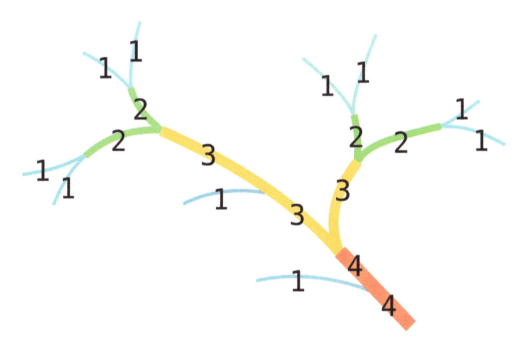

Figure 4.33: Strahler order method (Image from
https://commons.wikimedia.org/wiki/File:Flussordnung_(Strahler).svg, License CC BY-SA 3.0)

1. Search for Strahler in the *Processing Toolbox* and select SAGA | Terrain Analysis - Channels | Strahler order (figure 4.34).

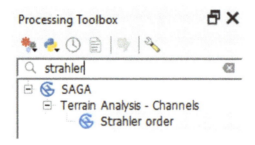

Figure 4.34: Strahler order tool

2. In the dialogue, select dem_fill for the *Elevation,* use strahler.sdat as the *Strahler Order* output filename, and click *Run.* Click *Close* when the algorithm is done (figure 4.35).

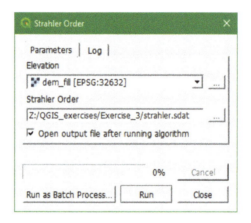

Figure 4.35: Strahler order dialogue

To make more sense of the strahler layer we are going to style it.

- Is the Strahler order layer a Boolean, discrete, or continuous raster?

The Strahler order layer is a discrete raster, but the order of the classes is important. For discrete rasters in QGIS we use the Paletted/Unique values styling. The higher the Strahler order, the bigger the stream. So we will use a color ramp from white to blue.

3. Open the *Layer Styling* panel if you closed it before and make sure that the strahler layer is selected.

4. Choose *Paletted/Unique values* from the drop-down menu.

5. Click *Classify* (figure 4.36, on the facing page).

To each unique value in the raster a unique random color has been assigned.

6. Right-click on *Random colors* and choose *Blues* (figure 4.37, on the next page).

The strahler layer now shows in an intuitive way that the higher orders will be larger streams than the lower orders (figure 4.38, on page 94).

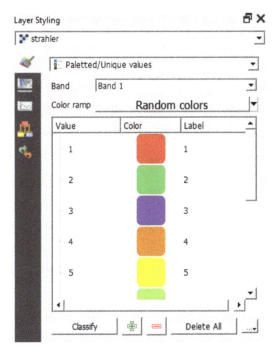

Figure 4.36: Styling of Strahler layer

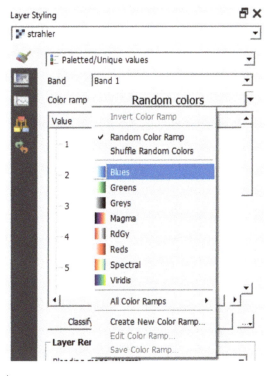

Figure 4.37: Select blues color ramp

Calibrate Strahler Threshold to Determine Streams

The next step is to apply a calibration procedure to determine which Strahler orders we consider to be streams. We will create Boolean layers for Strahler orders larger than or equal to

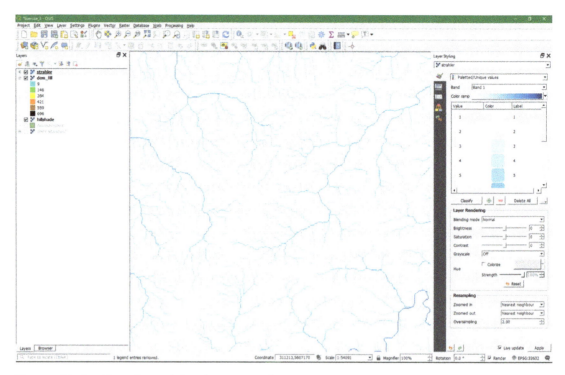

Figure 4.38: Strahler orders styled with intuitive colors

a threshold value. Each Boolean layer will be compared with a reference layer. Here we will compare the Strahler orders with the rivers on OpenStreetMap. For areas where there is a lack of data in OpenStreetMap, you could use the Google Satellite. Both OpenStreetMap and Google Satellite layers can be found in the *QuickMapServices* plugin.

7. Use the *Raster Calculator* to create a Boolean map with 1 (True) for Strahler order >= 5 and 0 (False) for the other values. Go to `Raster | Raster Calculator`.

8. Fill in the dialogue as in figure 4.39, on the next page. Call the output file `strahler5.tif` and click *OK* to run the calculation.

It is also good practice to style the `strahler5` layer.

9. Go to the *Layer Styling* panel and make sure that `strahler5` is selected.

10. Choose *Paletted/Unique values* from the drop-down menu.

11. Click *Classify*.

This layer is Boolean and therefore it has only pixels with values 0 and 1. For our calibration it is important to make the zeros transparent and the ones blue so we can compare the raster with the streams on OpenStreetMap.

12. Click on the color for value 1 and change it to dark blue. Don't bother about the color for value 0, because we'll make those pixels transparent (figure 4.40, on the facing page). You can add labels `stream/no stream`.

13. To make the zeros transparent go to the transparency tab and add 0 as *additional no data value* (figure 4.41, on page 96).

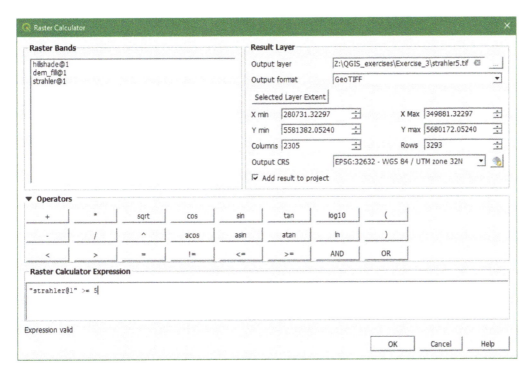

Figure 4.39: Calculate a Boolean map with the Raster Calculator

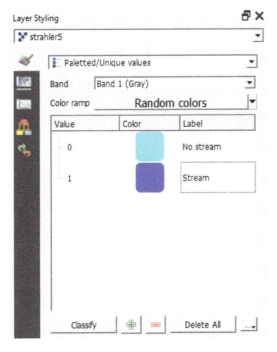

Figure 4.40: Styling a Boolean layer

In the Map Canvas we can see now all streams larger than or equal to Strahler order 5 and we can compare them with the rivers on the OpenStreetMap.

14. Repeat the steps in this subsection for different Strahler order threshold values and determine the one that corresponds best with the rivers on OpenStreetMap. Remove the other

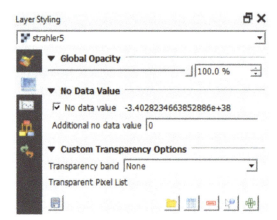

Figure 4.41: Add Boolean 0 as no data to make it transparent

Boolean layers.

> Tip: You can copy the styles of the layers: right-click on a layer, choose Styles | Copy Style and then right-click on the target layer and choose Styles | Paste Style (figure 4.42). You still need to make the zeros transparent, because that can not be copied.

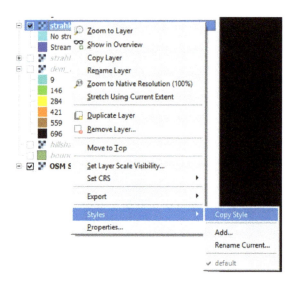

Figure 4.42: Copy the style of a layer

Calculate Channel Network, Flow Direction and Basins

The next step is to calculate the channel network, based on the Strahler threshold determined in the previous step. With the same tool we can also calculate the flow direction layer and a layer with all basins.

> In the *Processing Toolbox* you can also find related tools that use different algorithms. Always check the documentation of the algorithm to find the right one or do a comparative analysis.

15. Search for `Channel` in the *Processing Toolbox* and select `SAGA | Terrain Analysis - Channels | Channel network and drainage basin` (figure 4.43).

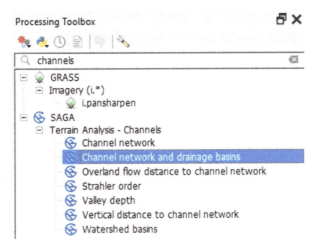

Figure 4.43: Search for channels in the Processing toolbox

16. In the dialogue, make sure that you select the `dem_fill` layer for the elevation and put the threshold at the value you found during the calibration in the previous section, e.g. 8. This means that streams with a Strahler order equal to or larger than this value will be considered as channels. The algorithm will calculate flow direction and the Strahler order to determine the channels and drainage basins. Select the checkboxes for the following outputs:

- Flow direction and save it as `flowdir.sdat`

- Channels and save it as `channels.shp`

- Drainage basins and save it as `basins.shp` (the other one is for raster)

17. Once the dialogue is setup similar to figure 4.44, on the next page, click *Run*.

> A flow direction layer indicates the direction of flow for each pixel. Direction can be expressed as compass direction, however we can not store text in a raster. Compass direction can also be expressed by degrees on a circle where north is 0 degrees, east is 45 degrees, etc. To store 360 degrees we would need more than 8 bits (0-255), which would increase the file size. In addition the D8 method uses discrete directions to the surrounding pixels. Therefore GIS software recodes the 8 directions. Each software, however, does it in their own way. The SAGA algorithm used in the QGIS Processing Toolbox uses 0 for north, 1 for northeast, etc. Which means it stores 0 to 7. A value of 255 is used for flat surfaces.

Styling and Inspecting the Results

Next you will style the output layers to get a better understanding of the results.

18. Open the *Layer Styling panel* and set the target layer to *channels*.

19. Choose the *Graduated* renderer with the following settings:

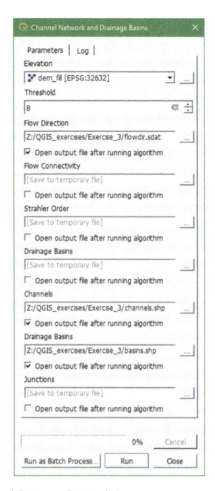

Figure 4.44: Channel Network and Drainage Basins dialogue

- Set the *Column* to ORDER and the *Precision* to 0 since Strahler Order values are integers.
- Set the *Method* to *Size*.
- Set the *Size from* values from 0.3 to 1.0.
- Keep the default *Mode* as *Equal Interval* with the same number of Strahler Order classes that exist in your data. Review the attribute table if need be. In this example it is set to 3 classes. Click *Classify*.
- Set the Legend values to integer values if need be.

20. Click the *Change* button and set the *Color* to an RGB value of 15 | 66 | 220.

21. Click the *Go back* ◀ button to return to the main symbology panel (figure 4.45, on the facing page).

22. Your map should now resemble that in figure 4.46, on the next page, showing the streams identified as having higher Strahler Orders are the main channels and the smaller ordered streams are tributaries.

Next you will work on styling the flow direction raster. This will help you check the results and ensure they match what you were expecting. You will employ a circular color ramp to

Figure 4.45: Styling Channels

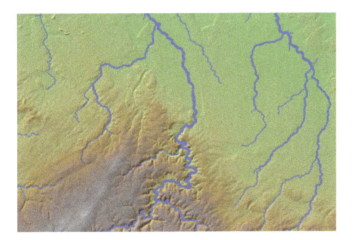

Figure 4.46: Channels Data Styled by Strahler Order

intuitively show southern oriented flows with warm colors and northern facing flows with cool colors.

23. Using the *Layer Styling panel* set the target layer to *flowdir*.

24. Set the renderer to *Paletted/Unique values* and click *Classify*.

25. Since the value of 255 represents flat surfaces, select that value and click the *Minus* button to remove it from the classification.

26. Choose the *Spectral* color ramp as a beginning point. Right-click on the color ramp and choose *Edit Color Ramp*.

27. The *Select color ramp* window will open. Here you can edit the color *stops* for the ramp. The idea is that the far left and right sides will represent northern flows and you will set both of those stops to the same blue color. You will then set the center stop to a bright yellow for

southern flows. The stop in the middle on the right will be western flows and the stop in the middle to the left will be eastern flows (figure 4.47).

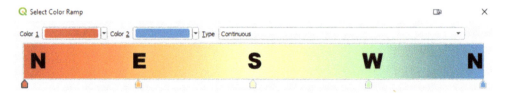

Figure 4.47: Beginning Circular Color Ramp

28. Click the drop-down menu for *Color 1* and choose *Pick color* from the context menu. With the eye dropper cursor click on the *Color 2*.

29. Click the center stop that will represent southern flows and assign it a bright yellow by entering these RGB values: 255 | 255 | 0.

30. For the eastern stop assign the color green by entering these RGB values: 0 | 255 | 0.

31. For the western stop use a magenta color, RGB: 214 | 60 | 170 (figure 4.48).

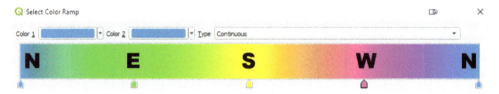

Figure 4.48: Final Circular Color Ramp

32. Click *OK* to accept the new color ramp.

33. Click the *Plus* ⊞ button to add a new value to the classification for the *Value* type 255. This represents the flat areas you initially removed before creating the circular color ramp. Click the color patch and choose white.

34. To review the *flowdir* data you will now give it a *Blending mode* of *Multiply* and turn on the *hillshade* layer. This will allow you to see the detail of the hillshade through the *flowdir* layer. At this point just have the *channels*, *flowdir*, and *hillshade* layers turned on. Zoom in to some of the more mountainous terrain in the south of the study area where there are also channels present. Inspect the data.

You should see southern facing slopes being represented by warmer colors and northern by cooler colors, eastern flows will be greener and western flows more purple (figure 4.49, on the facing page).

4.8   Define Outflow Point

A catchment is an extent or an area of land where surface water from rain, melting snow, or ice converges to a single point at a lower elevation, usually the exit of the basin, where the waters join another water body, such as a river, lake, reservoir, estuary, wetland, sea, or ocean. In order to delineate a catchment we need to have:

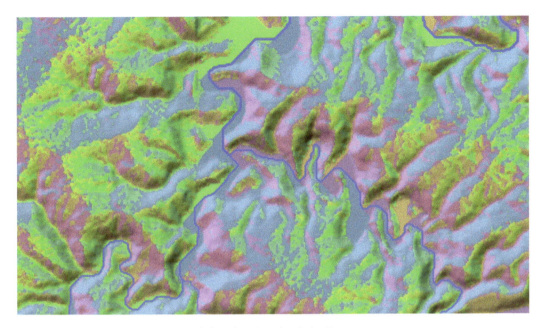

Figure 4.49: Flow Direction Raster Styled with a Circular Color Ramp

- the coordinates of our outlet in the same coordinate system as the map we are using

- the channel network that matches the flow directions as calculated from a hydrologically correct DEM

The outflow point of the Rur catchment is in Roermond, where the Rur enters the Meuse river (Maas in Dutch). The channel network that has been derived in the previous step is in the channels layer.

1. Make sure you have the channels layer on top of the OSM Standard layer from the *QuickMapServices* plugin.

2. Look for the location where the Rur river flows into the Meuse. Note that on the OpenStreetMap the Dutch names are used, because it lies in the Netherlands. Rur is spelled as Roer and Meuse is spelled as Maas.

> Note that the delineated channels are not corresponding well with the channels on OpenStreetMap. This can be for the following reasons: (1) Incorrect automatic delineation of streams, which can be caused by errors in the DEM or areas that are too flat, (2) Distortion due to (on-the-fly) reprojection and (3) Human influence on the natural course of the channels. The catchment delineation, however, only works when the outlet is defined on a delineated channel, because that corresponds with the filled DEM.

3. Choose a point on the delineated channel that is close to the real outlet of the Rur in the Meuse (step 2).

4. Use the *Coordinate Capture* tool to get the coordinates of the outflow point on the channels layer. You can find the *Coordinate Capture* tool from the main menu: Vector | Coordinate capture. Note that the plugin needs to be installed and activated (check the box in the *Plugins Manager*). The *Coordinate Capture* panel will appear (figure 4.50, on the next page).

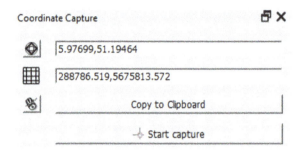

Figure 4.50: The Coordinate Capture panel

5. Click on *Start capture* and click on a point on the delineated river close to the mouth of the Rur as indicated in figure 4.51. Make sure you capture it from the river in the channels layer, because the background map (OpenStreetMap) has not been derived from the DEM and will result in errors in the catchment delineation.

Figure 4.51: Choosing the outflow point on the delineated channel

- What projection are we using? How do you know?

4.9 Delineate the Catchment

Now we're going to use these coordinates to calculate the upstream area (i.e. catchment) that produces discharge at this point.

1. In the *Processing Toolbox* search for `Upslope Area` and choose `SAGA | Terrain Analysis - Hydrology | Upslope Area` (figure 4.52).

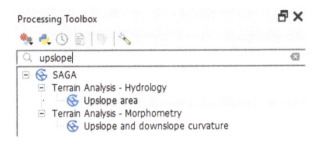

Figure 4.52: Upslope area tool in the Processing toolbox

2. Fill in the dialogue as in figure 4.53. Use the coordinates that you captured near the outflow point. For *Elevation* use the `dem_fill` layer. Use the default D8 method. Name the output *Upslope Area* `catchment.sdat`. Click *Run*. Click *Close* after the algorithm is done.

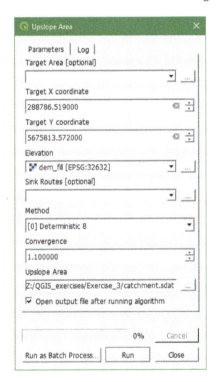

Figure 4.53: Upslope area tool dialogue

The result should look like the screenshot in figure 4.54, on the next page if you zoom to the extent of the layer.

In order to overlay the catchment boundary with other data, it is better to convert it from raster to vector (polygon).

3. To convert the raster layer to vector, go to the main menu and choose `Raster | Conversion | Polygonize (Raster to vector)` (figure 4.55, on the following page).

4. Make sure you choose the right input and call the output `Rur_catchment.shp`. Click *Run*. Click *Close* to get back to the main screen (figure 4.56, on page 105).

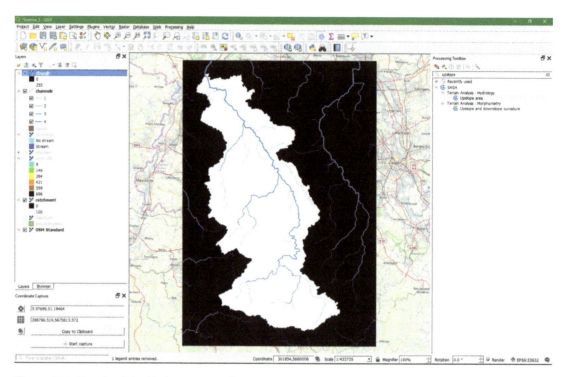

Figure 4.54: Result of the Upslope area tool

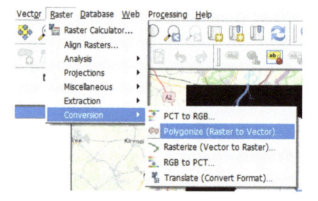

Figure 4.55: Polygonize menu

5. Look at the result. Also check the attribute table (right-click on layer name and choose Open
attribute table).

- What value is assigned to cells inside the catchment polygon? What value to those outside?

In the *Upslope Area* calculation, cells belonging to the catchment get a value of 100, while the other
cells get a value of 0. During the conversion to polygons, it can happen that geometry errors are
introduced. If you find more than one feature with a value of 100 this indicates a geometry error
(incorrect topology), because the boundary of the polygon makes a loop. See figure 4.57, on the
facing page. This can give errors when we use the polygon for geoprocessing.

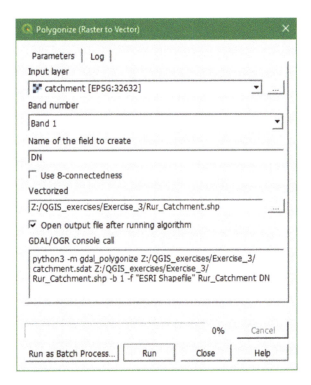

Figure 4.56: Polygonize dialogue

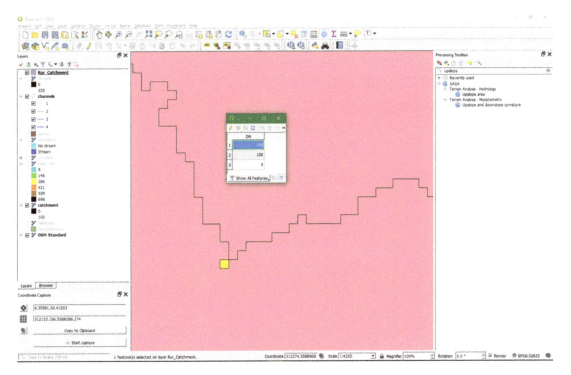

Figure 4.57: Geometry error caused by polygonizing a raster

Of course we are only interested in the catchment area, so we have to remove the outside polygon.

6. In the attribute table, toggle to editing mode using the ✏️ button, then select the record that you want to remove by clicking on the row. The selection will be highlighted in yellow on the map. Click the 🗑️ button to delete the selected feature, toggle off editing by clicking ✏️ again, and save the changes (figure 4.58).

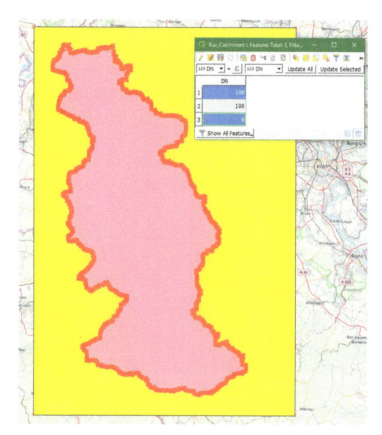

Figure 4.58: Delete polygons with value of 0

7. Now remove all unnecessary layers from the layers list so that we have only channels, Rur_catchment, dem_fill, and OSM Standard (in that order).

For visualisation it is nicer to clip the layers to the boundary of the catchment.

8. Let's first clip the channels vector layer to see only the streams that are inside the catchment. Go to the main menu and select Vector | Geoprocessing Tools | Clip (figure 4.59, on the facing page).

9. Fill in the dialogue as in figure 4.60, on the next page to use the catchment layer as a "cookie cutter" to clip the *Input layer* channels to the boundary of the *Overlay layer* Rur_catchment. Call the *Clipped* layer channels_clip.shp. Click *Run* to run the tool. Click *Close* to return the main screen.

> If you get an error related to the geometry, you can try to calculate a buffer of 0 meters to correct it. Go to the main menu and choose Vector | Geoprocessing Tools | Buffer. Another solution is to try tools from the *Processing Toolbox* that are less sensitive to geometric errors.

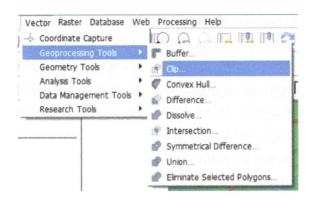

Figure 4.59: Clip vectors menu

Figure 4.60: Clip vectors dialogue

10. Remove channels from the layers list and drag the channels_clip layer to the top of the list.

11. Clip the DEM in the same way as we did in the section "Subset DEM", on page 82. The only difference is that we have to give a no data value that is out of the range of elevations. You can use -9999.

4.10 Storing the Data in a GeoPackage

To keep the data together and enable easy distribution it is good to save the layers as a GeoPackage.

1. In the processing toolbox look for the *Package Layers* tool (figure 4.61, on the next page).

2. In the *Package Layers* dialogue click the ⎡...⎤ and select all layers. Note that these are only the vector layers (figure 4.62, on the following page).

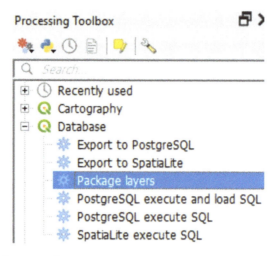

Figure 4.61: The Package Layers tool

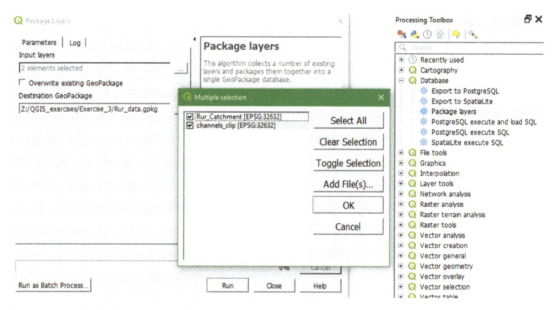

Figure 4.62: The Package Layers dialogue

3. Save it as `Rur_data.gpkg` and click *Run* and *Close* when it's done.

4. We can add the raster layers from the *Browser* panel. Simply drag the raster layers (e.g. `dem_catchment`) to the `Rur_data.gpkg`. You might need to refresh the *Browser* panel to see the GeoPackage.

4.11   Styling the Resulting Catchment Area

To show the results of your analysis you can use a technique named *Inverted Polygon Shapeburst Fills* to focus attention on the Rur Catchment.

1. Open the *Layer Styling panel* by clicking the 🖌 button. Set the target layer to *Rur_Catchment*.

2. Change from a *Single symbol* renderer to an *Inverted polygons* renderer. This renders the data as the inverse of it's geometry. This creates a mask around the Rur valley.

3. Next select the *Simple Fill* component. Then change the *Symbol layer type* to *Shapeburst fill*. In the *Gradient colors* section use the default *Two color* method. Change the first color to an RGB value of 135 | 135 | 135. Change the second color to white.

4. In the *Shading style* section, click the *Set distance* option and set the distance to 4.

5. Under *Layer rendering* set the *Opacity* to 65%.

6. Finally at the top of the Layer Styling Panel select the *Shapeburst fill* component.

7. Click the *Add symbol layer* ⊕ button. Change the new *Simple fill* renderer to a *Symbol layer type* of *Outline simple line*. Give it a *Color* of black and a *Stroke width* of 0.46.

This gives us the nicely styled basin as shown in figure 4.63.

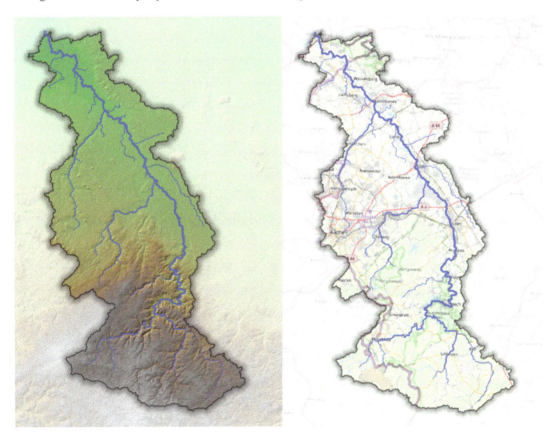

Figure 4.63: Rur Basin Highlighted

The following video explains the steps in this chapter: "Stream and Catchment Delineation in QGIS 3". If you have voids in your DEM you can watch "Interpolate Voids in DEM for Catchment Delineation in QGIS". Videos can be found at the YouTube channel http://www.youtube.com/c/hansvanderkwast.

5. Adding Open Data to Your Catchment

5.1 Introduction

Now that we have defined the boundaries of our study area, the Rur catchment, we can look for open access data that is available on the Internet.

After this chapter you will be able to:

- add *OGC web map services* to QGIS

- download vectors from *OpenStreetMap* using the *QuickOSM plugin*

- style open data from the web

In this exercise we will use:

- Data from the European Environment Agency (EEA) public map services: `http://discomap.eea.europa.eu`.

- Data from OpenStreetMap: `http://www.openstreetmap.org`

Instead of using a web browser, we will create a live link between the online data and our data in QGIS, using Web Map Services (WMS), which is an open standard for sharing maps through the Internet. We will also download vector layers from OpenStreetMap through QGIS.

For this exercise you need the delineation of the Rur catchment and its streams from the previous exercise.

Before we start, make sure that you have the following layers loaded into QGIS:

- Boundary of the Rur catchment

- Inverted polygon shapeburst fill mask layer from the previous chapter

- Streams in the Rur catchment

- OSM Standard backdrop from the QuickMapServices

There are several ways to begin this new map document while maintaining the symbology established during the catchment delineation exercise. You can open the map document from the previous chapter and choose Save As. You can also open a new map document and copy/paste the layers from the catchment delineation exercise into it. To do this have the project open from the previous chapter, select these layers, right-click and choose Copy Layer from the context menu. Then right-click in the Layers Panel for the new project and choose Paste Layer/Group.

Your map should now look like figure 5.1.

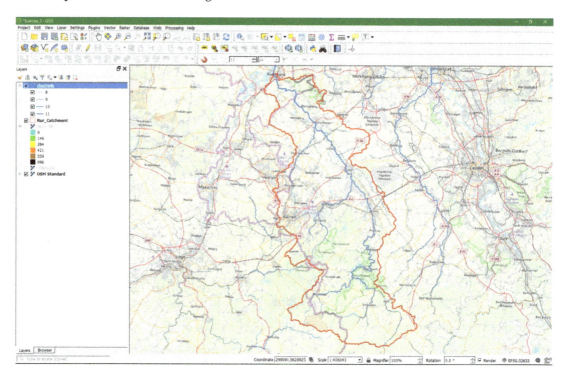

Figure 5.1: The map at the start of Chapter 6

5.2 Adding Data from Web Map Services

In this section we are going to use data from a web map service. In this example we will use data from the European Environment Agency (EEA). The EEA is an agency of the European Union (EU) that provides independent information on the environment, thereby helping those involved in developing, adopting, implementing, and evaluating environmental policy, as well as informing the general public. The European Environment Agency provides maps on thematic areas such as air, water, climate change, biodiversity, land, and noise. The map services are accessible through `http://discomap.eea.europa.eu`.

1. Go to the website `http://discomap.eea.europa.eu`. There you will see a list of themes.

2. Click on *Land* under *Land*. Here you can see the EEA datasets related to land cover and their descriptions.

3. Look for the CLC2012 Web Mercator (MapServer) data and click on *More info* to see what data it contains.

4. Click on the picture of the map on the left to open an interactive map viewer.

5. In the map viewer click on the *Legend* tab and zoom in on the Rur catchment.

The screen now looks like figure 5.2.

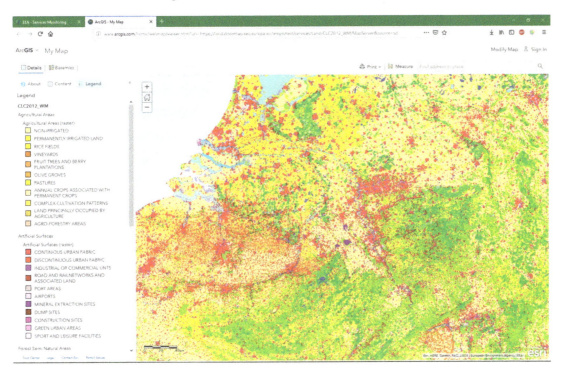

Figure 5.2: CORINE 2012 at the EEA portal

6. Now we're going to add this land cover data to our QGIS project. Go to your QGIS project and click the Open Data Source Manager button from the toolbar under the main menu.

7. In the *Data Source Manager* dialogue choose WMS/WMTS.

8. In the dialogue that opens click the *New* button.

9. In the dialogue that follows, type EEA CORINE 2012 for *Name*.

10. We can find the URL on the website below the title of the CLC2012 Web Mercator (MapServer). Right-click on WMS and choose Copy link location (figure 5.3, on the following page).

11. Paste the link in the *Create a New WMS/WMTS Connection* dialogue in QGIS (figure 5.4, on the next page). Keep the defaults and click *OK*.

12. Back in the other dialogue window click *Connect*. The layers will now be retrieved from the WMS server (figure 5.5, on page 115).

13. Click on Wetlands so that it is highlighted, choose *PNG* for the *Image encoding format* (this allows transparency), and check the box before *Use contextual WMS legend*. Then click *Add* and *Close* to return to the main screen. Note that WMS layers are georeferenced pictures, not vectors.

Figure 5.3: Copy link location

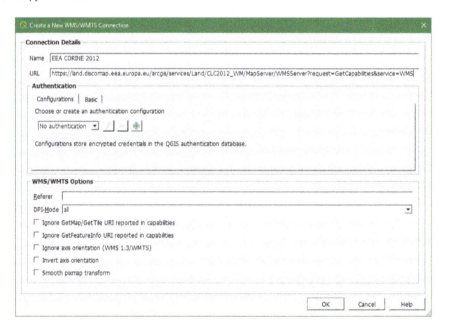

Figure 5.4: Create new WMS connection

- You can see that the source area of the Rur catchment is located in wetlands. What are the names of these wetlands?

- In which country/countries are these wetlands?

14. In a similar way, add Artificial Surfaces.

- In the midstream part of the catchment there are some large purple areas. What are these?

- Add *Google satellite* from the *QuickMapServices* to have a closer look at these features. Describe what you see.

- What are the other Artificial Surface classes in the map? Can you describe the geographical distribution of population in the Rur catchment?

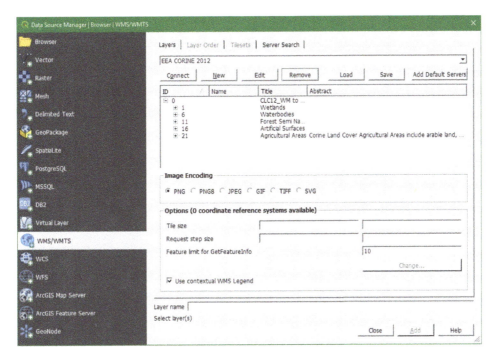

Figure 5.5: WMS connection

- What are the largest cities in the catchment?

15. Load the remaining classes.

- Where are the forests and agricultural areas located?

16. Now rearrange the layers so that you have a map with the catchment boundary, stream network, and land cover classes (figure 5.6, on the next page).

17. Save the project before you continue.

If you have time, you can look for more EEA data in a similar way or look for other web map services.

> There are different web services you can use in QGIS. There are the OGC services: WMS, WFS, and WCS. While WMS renders a picture from the data, WFS and WCS will give you the vector and raster data respectively. You can also connect to Spatial Data Infrastructures that use GeoNode or to AcGIS Map Server and ArcGIS Feature Server. These connections are available in the *Data Source Manager* and the *Browser* panel.

5.3   Adding Vector Data from OpenStreetMap

OpenStreetMap (OSM) is a collaborative project to create a free editable map of the world. OSM is considered a prominent example of volunteered geographic information (VGI) or crowd-sourcing. There are several ways to use the data:

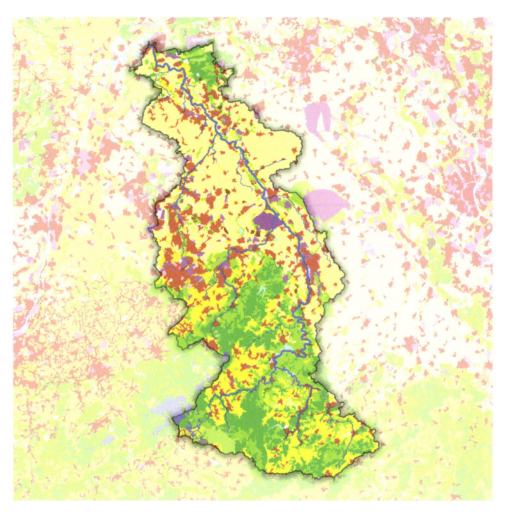

Figure 5.6: WMS combined with other layers

- Through the interactive map on the OpenStreetMap website (http://www.openstreetmap.org)

- In QGIS you can add OSM basemaps via the *QuickMapServices* plugin

- In QGIS you can download OSM data directly over the Internet. There are several ways to do this. In this section we'll use the *QuickOSM* plugin which uses the Overpass API (https://wiki.openstreetmap.org/wiki/Overpass_API).

In this section we are going to download OSM vector data directly in QGIS for the Rur Catchment. We continue from the previous results, but we only visualise the Rur_Catchment and Channels_Clip layers. The other layers should be unchecked.

1. Install the *QuickOSM* plugin through the main menu: Plugins | Manage and Install Plugins. Search for QuickOSM.

2. Open the *QuickOSM* dialogue by choosing Vector | QuickOSM | QuickOSM from the main menu (figure 5.7, on the facing page).

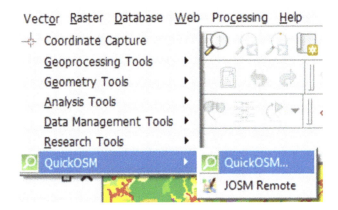

Figure 5.7: QuickOSM menu

We're first going to download the rivers so we can compare them with the rivers that we have previously derived. The OSM data attributes consist of keys and values. To learn more about it, click *Help with key/value*.

3. Choose *waterway* as *Key*, *river* as *Value*. Choose the Rur_catchment polygon as the extent. Note that you can also select the extent of the Map Canvas. Click the arrow before *Advanced* and make sure only *Node*, *Way*, *Relation*, and *Lines* are checked (you need to select the geometry that you would like to download). The dialogue should now look like figure 5.8. Click *Run Query*.

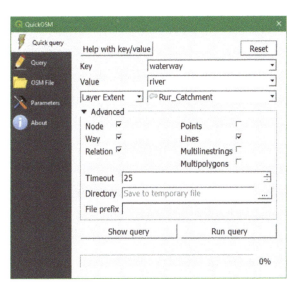

Figure 5.8: QuickOSM dialogue

> You can increase the *Timeout* value if you have a slow Internet connection.

The new layer will be added as a temporary scratch layer as indicated by the Temporary Scratch Layer icon in the indicator space to the right of the layer in the Layers Panel.

4. Adjust the style and compare the waterway_river from OSM with the Channels_Clip layer.

- What do you observe?

5. Let's add the mines in a similar way. Use key=landuse and value=quarry. Don't forget to select *Multipolygons* instead of *Lines*.

6. Style the polygons with a gray fill and black stroke using an *italic* font.

7. Label the polygons with the Name attribute. Switch to the label *Formatting* tab ⁺ab and enter a space as the *Wrap on character*. Then set the *Alignment* to *Center*. Switch to the label *Rendering* tab 🖋 and click *Only draw labels which fit completely within feature*.

8. Try to locate Jülich. You can also use the *GeoCoding* plugin to locate Jülich. You can install it through the main menu: Plugins | Manage and Install Plugins. Search for GeoCoding. The *GeoCoding* plugin uses web services (Nominatim and Google) to retrieve the coordinates of an address.

9. Now zoom in on the centre of Jülich (see figure 5.9).

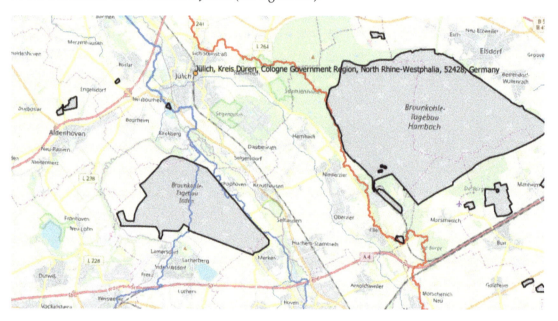

Figure 5.9: The area around Juelich (Germany)

Southeast of Jülich is the Forschungszentrum Jülich, a large research institute. In the south and east we see a large surface mining of lignite. The one in the south is in the Rur catchment.

10. Now compare the OSM derived quarry with a Google Satellite and the EEA artificial surfaces map.

- What are the differences?

- Which one is more up-to-date? Google Satellite, EEA, or OSM?

- How does the hydrography relate to the quarry (OSM versus GIS delineated)?

Remember that the layers that were added via *QuickOSM* are temporary scratch layers, indicated by ⌣ . There is an option to make the layer permanent, but the export layer option is more flexible. We will use that to save the layer to the GeoPackage created in the previous chapter ("Storing the Data in a GeoPackage", on page 107).

11. Right-click on the layer, choose Export | Save Features As (figure 5.10).

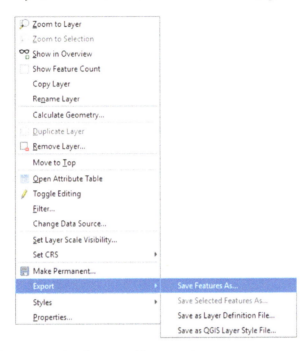

Figure 5.10: Export a layer to another format and/or projection

12. In the *Save vector layer as* dialogue (figure 5.11), for *Format,* choose *GeoPackage*. At *Filename* browse to Rur_data.gpkg created in the previous chapter. At *Layer name* type Quarries. This will be the name of the layer inside the GeoPackage. Change the *CRS* to the one of the project (EPSG: 32632). Click *OK*.

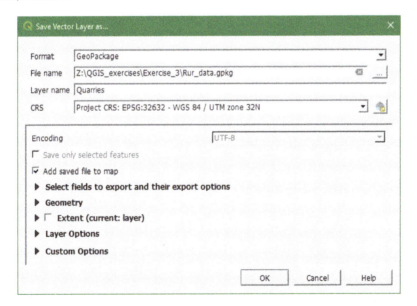

Figure 5.11: Save a layer to a GeoPackage

13. Now add a few other interesting features (points, lines, and polygons) and add them to the GeoPackage:

- Dams: Key=waterway, value=dam

- Lakes: Key=natural, value=water

- Springs: Key=natural, value=spring

14. Now style the layers. Begin with the lakes. You can use the same symbology and label settings used in the "Preparing Data from Hardcopy Maps" chapter, on page 13. The simplest way to achieve this is to open that map document, right-click on the lakes layer, and choose *Styles | Copy Style | All Style Categories* from the context menu. Then bring up the current map document and again right-click on the lakes layer and choose *Styles | Paste Style | All Style Categories* from the context menu. Switch to the label *Rendering* tab ✎ and click *Only draw labels which fit completely within feature.*

Another option is to open *Layer Properties* for the lakes layer in the map document from the "Preparing Data from Hardcopy Maps" chapter. Switch to the *Symbology* tab and expand the *Layer Rendering* section. Click the *Style* menu and choose *Save style*. In the *Save Layer Style* window save the style as lakes.qml in the exercise folder. Switch to the current project. Open *Layer Properties* for the lakes layer and in the *Layer Rendering* section, click the *Style* menu and choose *Load style*. Choose the lakes.qml file just saved. This is a good option if you will be reusing a style repeatedly.

15. Next you will work with the dam lines. Give them a *Color* of black and a *Stroke width* of 0.86.

16. Click the *Add symbol layer* button ⊞ . Select the *Simple line* component and choose a *Symbol layer type* of *Marker line*. Select the *Simple marker* component and choose the vertical line symbol from the choices displayed below. Increase the *Stroke width* to 0.2 and the *Size* to 3. Your dams symbol preview should now look like this ┼┼┼┼┼┼┼ .

17. Finally you will style the spring points. Make the layer the target layer in the *Layer Styling Panel*. Select the *Simple fill* component. Choose a *Symbol layer type* of *SVG Marker*. Select the *symbol* folder and find the *blue-marker.svg* 📍 . Increase the *Size* (Width and Height) to 6 each.

> The following video explains the steps in this chapter: "Adding Open Data to your QGIS Project". Videos can be found at the YouTube channel http://www.youtube.com/c/hansvanderkwast.

6. Calculating Percentage of Land Cover per Subcatchment

6.1 Introduction

For studies on catchment hydrology it is often important to know the percentage of land cover per subcatchment.

After this chapter you will be able to:

- add unique values to vector layers
- calculate areas of polygons in attribute tables
- use a *SpatiaLite database*
- style a vector layer using a `.qml` file
- *clip*, *reproject*, and *export* vector layers to another format
- use *conditions* in the *field calculator*
- apply vector *geoprocessing tools* such as *dissolve*, *intersection*, and *buffer*
- correct geometry errors
- create pie charts using the *Data Plotly* plugin

In this chapter we'll use:

- `catchpolygons.shp`: shapefile with subcatchments
- CORINE 2018 Land cover

The first file has been created by delineating subcatchments within the Rur using the methods described in the chapter "Stream and Catchment Delineation", on page 71.

The data can be downloaded from `http://locatepress.com/qgis_hydrological`.

Figure 6.1, on the following page shows the workflow for this chapter.

6.2 Preparing the Subcatchment Layer

1. Download the `catchpolygons.shp` shapefile and add it to an empty QGIS project.

2. Open the attribute table of the `catchpolygons` layer.

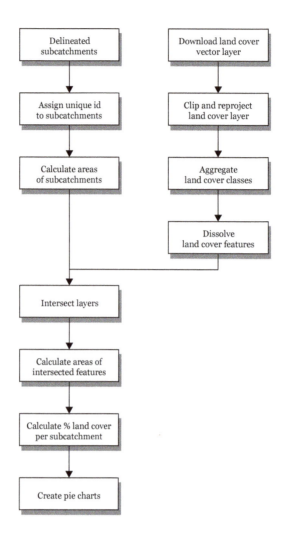

Figure 6.1: Workflow for calculating the percentage of land cover in each subcatchment

This layer was calculated using the catchment delineation approach from the chapter "Stream and Catchment Delineation", on page 71. The result is that each subcatchment has a value of 100. For our purpose, however, each subcatchment needs a unique id.

3. Toggle to editing mode using the ✎ button.

4. In the *Field Calculator* formulate the equation DN = $id and click *Update All* (figure 6.2, on the facing page). Click the 💾 button to save the edits.

The $id function assigns the unique feature id to each feature in the attribute table.

Now we need to add an attribute with the surface area for each subcatchment polygon.

5. Click 🗒 to add a column.

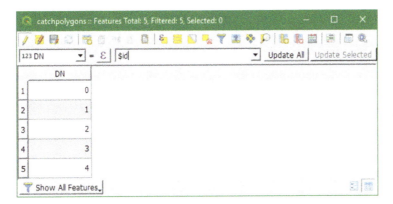

Figure 6.2: Number features using the unique feature id

6. Call the new field `CatchArea`, indicate that it contains `Decimal number (real)` with *Length* 10 and *Precision* 2 (figure 6.3). Click *OK* to create the column.

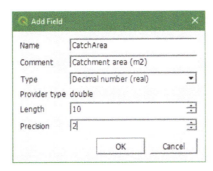

Figure 6.3: Add a field for subcatchment area

7. In the *Field Calculator* formulate the equation `CatchArea = $area` and click *Update All*.

The attribute table now has the unique id's and surface areas for each subcatchment (figure 6.4)

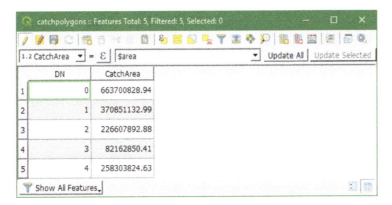

Figure 6.4: Attribute table with unique id's and areas for each subcatchment

8. Click the to save the edits and toggle editing off.

6.3   Preparing the Land Cover Data

Downloading CORINE 2018 Data

In the previous chapter we used the CORINE 2012 WMS data. For calculations, however, we can't use the WMS rendered pictures—we need access to the data. The CORINE 2018 data can be downloaded from the Internet.

1. Open a browser and go to `https://land.copernicus.eu/pan-european/corine-land-cover`.

2. Click on CLC 2018.

You'll see an online map with a preview of the layer.

3. Click the *Download* tab.

4. For downloading data you need an account. Click the link to register.

Once logged in you can choose from different data formats (figure 6.5).

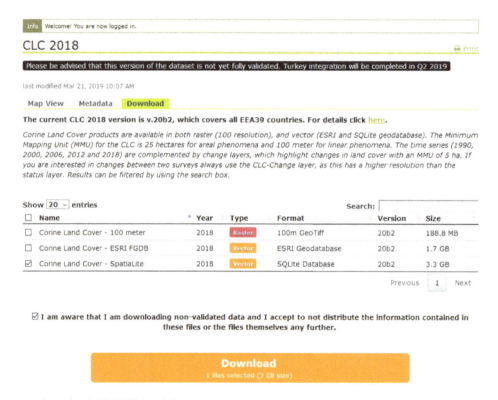

Figure 6.5: Download CORINE Land Cover

Here we'll use the SpatiaLite format, because we need vector data. SpatiaLite (an extension to `SQLite`) is an open spatial database format supported by QGIS.

5. Check the box before the SpatiaLite format and check the box that you are aware that this is non-validated data.

6. Click the *Download* button.

The file is 3GB. If you have low bandwidth, you can also download a subset of the dataset from the course materials at http://locatepress.com/qgis_hydrological.

7. Unzip the contents to a new folder that you'll use for this exercise.

8. Drag the SpatiaLite layer CLC2018_CLC2018_V2018_20b2 to the Map Canvas.

9. Make the polygons of the catchpolygons layer transparent with a black outline and stroke width of 0.66 mm.

Style the CORINE 2018 Land Cover Map

Now we're going to style the CORINE 2018 Land Cover map. In the downloaded dataset you'll find a Legend folder. The folder contains legend files in different formats. Here we'll use the clc_legend.qml file, a QGIS layer style file.

10. Right-click on the land cover layer and choose Properties. Go to the *Symbology* tab.

11. At the bottom of the dialogue go to Style | Load Style (figure 6.6).

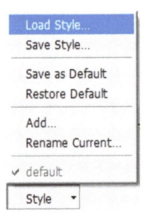

Figure 6.6: Load style

12. Now you're in the *Database styles manager* dialogue. Browse to the legend\clc_legend.qml file and click *Load style* (figure 6.7, on the next page).

The legend is now applied with the correct settings to the SpatiaLite layer (figure 6.8, on the following page).

13. Click *OK* to accept and return to the Map Canvas.

Clip and Reproject the Land Cover Map

The land cover layer is still too large (even if you downloaded the provided subset), so we're going to prepare a subset that only covers the subcatchments. In addition we're also going to reproject the land cover layer (EPSG: 3035) to the same projection as the catchpolygons layer (EPSG:32632). Both can be done with one tool.

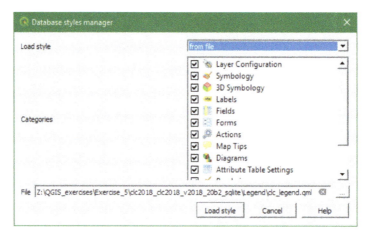

Figure 6.7: Database styles manager

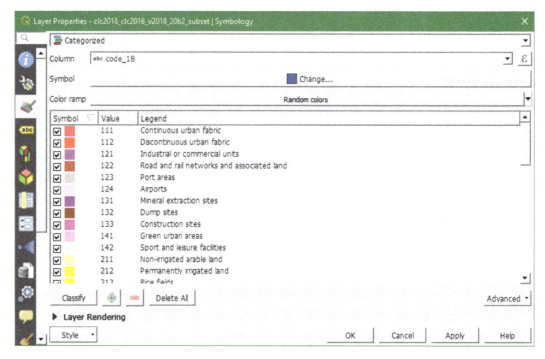

Figure 6.8: CORINE layer style

14. Right-click on the CLC2018_CLC2018_V2018_20b2 layer and choose Export | Save Features As.

15. Choose as a *Format ESRI Shapefile*, for *File name* corine2018_repr (make sure that you save it in the folder with the data for this chapter) and for *CRS* the CRS of the project (EPSG:32632).

16. Check the box before *Extent* and click *Calculate from Layer* and choose catchpolygons. Click *OK* (figure 6.9, on the next page)

17. Copy the style from the CLC2018_CLC2018_V2018_20b2 layer to the corine2018_repr layer and remove the CLC2018_CLC2018_V2018_20b2 layer.

The result should be like figure 6.10, on the facing page.

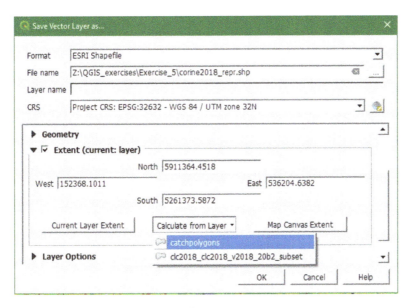

Figure 6.9: Export layer and change projection and extent

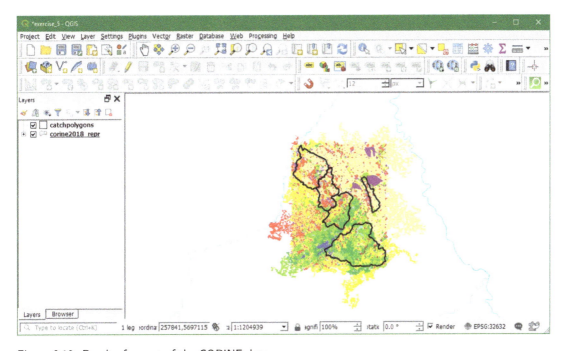

Figure 6.10: Result of export of the CORINE data

Aggregate Land Cover Classes

The CORINE data that we have is so-called level 3 data. In the attribute table the code_18 field has values for level 3, indicated by 3 digits. In this section we're going to aggregate the detailed level 3 classification to level 1. Therefore we need to create a new field with only the first digit for each feature.

18. Open the attribute table of the corine2018_repr layer and toggle to editing mode.

19. Add a new field. Give the attribute the *Name* `Level1`, choose for *Type* `Whole number (integer)` and *Length* 1 (figure 6.11).

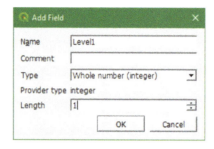

Figure 6.11: Add new field for CORINE Level 1

20. Use the Ɛ button to open the *Expression Dialog*.

We're going to write an expression that recodes all level 3 classes to level 1.

21. Write the expression as given in figure 6.12.

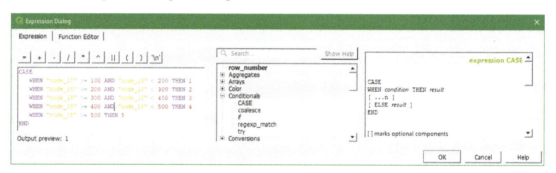

Figure 6.12: Expression dialog for recoding CORINE to level 1

22. Update the Level 1 field.

The attribute table should now look like figure 6.13, on the next page.

23. Toggle off the editing mode and save the edits.

Dissolve Land Cover Features

Now there are many contiguous polygons with the same level 1 class. These have to be merged into one feature. This is done with the *dissolve* operation.

24. In the main menu go to `Vector | Geoprocessing tools | Dissolve`.

25. In the *Dissolve* dialogue choose `corine2018_repr` as *Input layer* and select the *level1* field as the *Dissolve field*. Save the file as `corine2018_dissolved` and click *Run* (figure 6.14, on the facing page).

26. Click *Close* to return to the Map Canvas when the calculation is complete.

27. Open the attribute table of `corine2018_dissolved` to check the result. It should now show only 5 features with level 1 classes 1 to 5 (figure 6.15, on page 130). Note that you can click on the field name to sort the features.

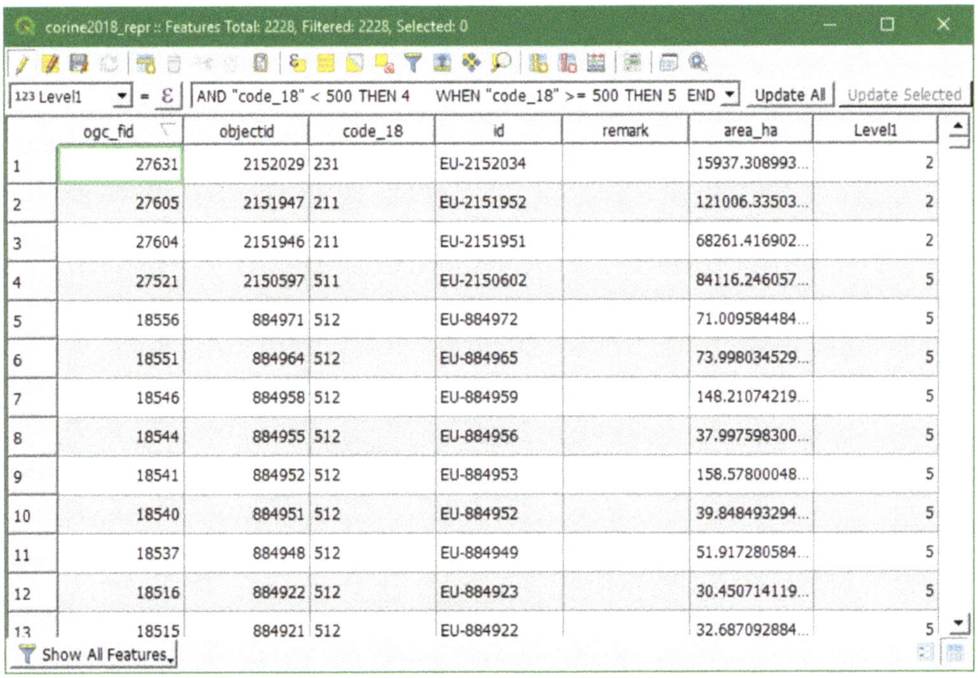

	ogc_fid	objectid	code_18	id	remark	area_ha	Level1
1	27631	2152029	231	EU-2152034		15937.308993...	2
2	27605	2151947	211	EU-2151952		121006.33503...	2
3	27604	2151946	211	EU-2151951		68261.416902...	2
4	27521	2150597	511	EU-2150602		84116.246057...	5
5	18556	884971	512	EU-884972		71.009584484...	5
6	18551	884964	512	EU-884965		73.998034529...	5
7	18546	884958	512	EU-884959		148.21074219...	5
8	18544	884955	512	EU-884956		37.997598300...	5
9	18541	884952	512	EU-884953		158.57800048...	5
10	18540	884951	512	EU-884952		39.848493294...	5
11	18537	884948	512	EU-884949		51.917280584...	5
12	18516	884922	512	EU-884923		30.450714119...	5
13	18515	884921	512	EU-884922		32.687092884...	5

Figure 6.13: Result of the recoding of CORINE to level 1

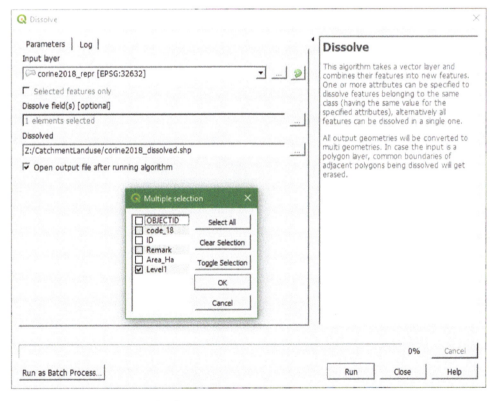

Figure 6.14: Dissolve CORINE level 1 features

28. Remove corine2018_repr from the *Layers Panel* and style the corine2018_dissolved layer with colors based on the original CORINE level 3 legend (figure 6.16, on the following page).

	OBJECTID	code_18	ID	Remark	Area_Ha	Level1	
1	5488	111	EU-5488		31.462886189...	1	
2	309985	211	EU-309985		194.84139484...	2	
3	1162914	311	EU-1162914		268.02058276...	3	
4	2054612	412	EU-2054612		37.868599070...	4	
5	2114525	511	EU-2114525		84116.246057...	5	

Figure 6.15: Attribute table CORINE 2018 after dissolving level 1 features

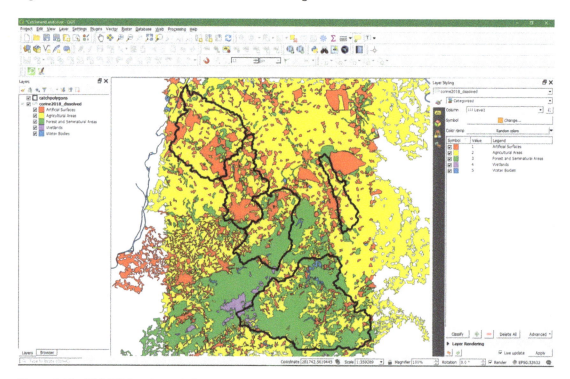

Figure 6.16: CORINE Level 1 styled

6.4   Intersect the Land Cover Layer with the Subcatchment Layer

Now that we have prepared the land cover and subcatchment polygon layers, the next step is to *intersect* the two layers to add the borders of the subcatchments to the land cover map.

1. In the main menu go to `Vector | Geoprocessing tools | Intersection` (figure 6.17, on the next page).

2. Choose the `corine2018_dissolved` layer as *Input layer*, choose `catchpolygons` as *Overlay layer*. Choose `corine_catch_intersected.shp` as output filename and click *Run* (figure 6.18, on the facing page).

You'll see the following error (figure 6.19, on page 132). This is related to geometry errors that can occur in the catchment delineation process where we polygonize a raster.

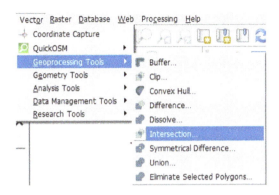

Figure 6.17: Intersection menu

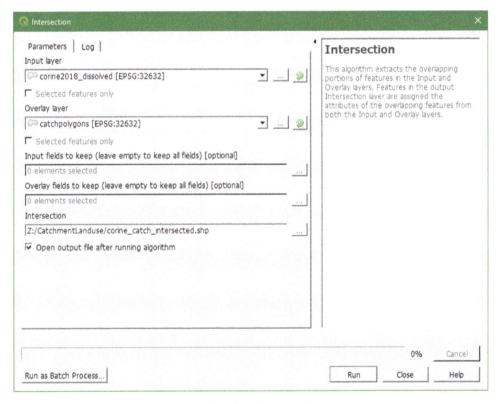

Figure 6.18: Intersect CORINE land cover with subcatchment polygons

The error can be corrected by calculating a *buffer* with a distance of zero meters. Buffers are normally used to create polygons at a fixed distance around features.

3. In the main menu go to `Vector | Geoprocessing tools | Buffer`.

4. In the *Buffer* dialogue choose `catchpolygons` as *Input layer* and set the buffer *Distance* to 0 meters. Keep the defaults and choose `catchpoly_buffer` as output filename and click *Run* (figure 6.20, on page 133).

5. Click *Close* after the processing is done.

6. Copy the style from `catchpolygons` layer to the `catchpoly_buffer` layer and remove the

Figure 6.19: Invalid geometry error

`catchpoly_buffer` layer.

7. Repeat steps 1 and 2 and call the output filename of the intersection layer `corine_catch_intersected_cor.shp`.

8. Style the `corine_catch_intersected_cor` layer by copying the style from the previously styled `corine2018_dissolved` layer. Check the result of the intersection.

9. Remove the `corine2018_dissolved` layer from the *Layers panel*.

6.5   Calculate Land Cover Class Area per Subcatchment

Now we need to calculate the CORINE level 1 class area per subcatchment.

1. Open the attribute table of the `corine_catch_intersected_cor` layer.

In the attribute table we can find the *CatchArea* field with the area of each subcatchment, the *DN* field with the unique ID for each subcatchment, and the *Level1* field with the Level 1 CORINE class for each feature. We'll add a new field and calculate the area of each feature which corresponds with the area of each Level 1 class in a subcatchment in a similar way as we have previously done for the subcatchment areas.

2. Toggle to editing mode.

3. Add a field `ClassArea` with *Type* as `Decimal number (real)`, *Length* 10, and *Precision* 2.

4. Use the *Field calculator* to calculate `ClassArea = $area` and click *Update all*.

5. Now add a field for the percentage of each CORINE Level 1 class in each subcatchment. Call the field `Percentage` with *Type* as `Decimal number (real)`, *Length* 4, and *Precision* 1.

6. Use the *Field calculator* to calculate `Percentage = (ClassArea/CatchArea) * 100`. You can

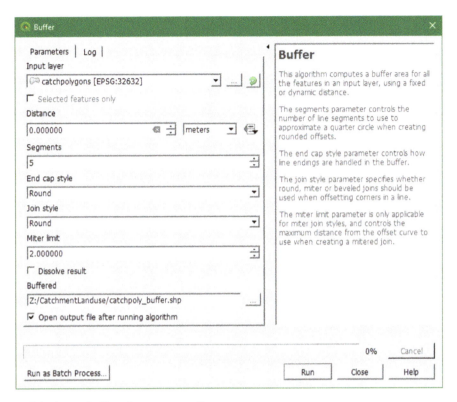

Figure 6.20: Calculate a buffer of zero meters distance to correct geometry errors

use the *Expression Dialog* to formulate the equation.

7. Click *Update all* to assign the percentage to each feature.

8. To complete the attribute table, add a field with the Level 1 class names as text. Call it `Landcover`. You can use the `CASE...WHEN` function for that again.

9. Toggle editing off and save the edits.

The attribute table should now look like figure 6.21.

corine_catch_intersected_cor :: Features Total: 21, Filtered: 21, Selected: 0

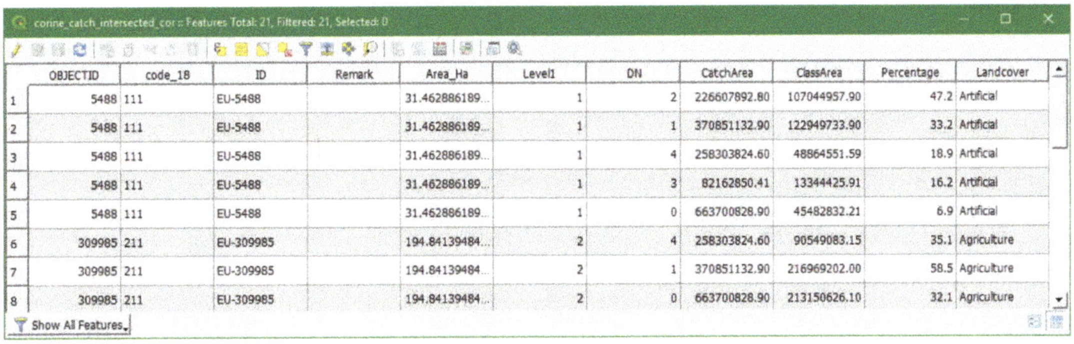

	OBJECTID	code_18	ID	Remark	Area_Ha	Level1	DN	CatchArea	ClassArea	Percentage	Landcover
1	5488	111	EU-5488		31.462886189...	1	2	226607892.80	107044957.90	47.2	Artifical
2	5488	111	EU-5488		31.462886189...	1	1	370851132.90	122949733.90	33.2	Artifical
3	5488	111	EU-5488		31.462886189...	1	4	258303824.60	48864551.59	18.9	Artifical
4	5488	111	EU-5488		31.462886189...	1	3	82162850.41	13344425.91	16.2	Artifical
5	5488	111	EU-5488		31.462886189...	1	0	663700828.90	45482832.21	6.9	Artifical
6	309985	211	EU-309985		194.84139484...	2	4	258303824.60	90549083.15	35.1	Agriculture
7	309985	211	EU-309985		194.84139484...	2	1	370851132.90	216969202.00	58.5	Agriculture
8	309985	211	EU-309985		194.84139484...	2	0	663700828.90	213150626.10	32.1	Agriculture

Show All Features

Figure 6.21: Attribute table after calculating the Corine Level 1 class area, percentage, and assigning class names

6.6 Create Pie Charts Using the Data Plotly Plugin

Now that we have the percentage of land cover for each subcatchment, we can create pie charts. One way is to copy the columns and paste it into spreadsheet software for further processing. Another way is to use the *Data Plotly* plugin, which we'll use here.

1. Install the *Data Plotly* plugin from the *Plugin Manager*.

We're going to make a pie chart for each subcatchment. We have previously assigned a unique ID for each subcatchment in the DN field.

2. From the attribute table of the corine_catch_intersected_cor layer, select all features with DN = 0. You can use the *Select features by expression* dialog by clicking the button. Alternatively you can sort the attribute table on DN and use the Ctrl button and select multiple features with DN = 0 (figure 6.22).

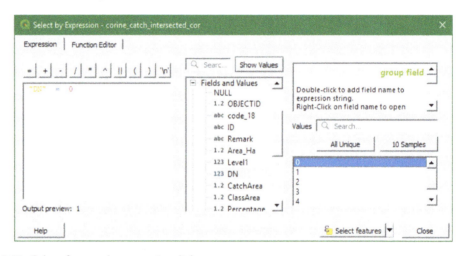

Figure 6.22: Select features by expression dialogue

3. Click on the DataPlotly button   to open the *DataPlotly* panel.

4. For *Plot Type* choose Pie Plot. Under *Plot Parameters* choose the corine_catch_intersected_cor layer, check the box before *Use only selected features*, for *Grouping field* choose Landcover, and for *Y Field* Percentage.

5. Click the   button to set the legend properties (figure 6.23, on the next page).

6. Click the *Create Plot* button and check the result.

7. With the *Export to image* button   you can export the plot to a .png file.

The result should look like figure 6.24, on the facing page.

> The following video explains the steps in this chapter: "Calculate the Percentage of Land Use per Subcatchment in QGIS3". Videos can be found at the YouTube channel http://www.youtube.com/c/hansvanderkwast.

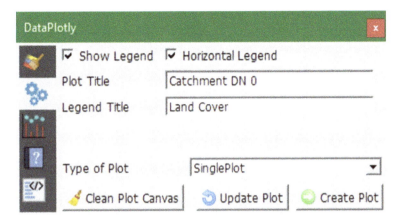

Figure 6.23: DataPlotly legend settings

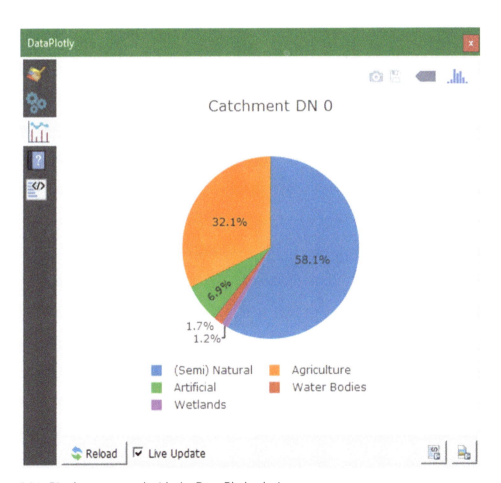

Figure 6.24: Pie chart generated with the Data Plotly plugin

7. Map Design

7.1 Introduction

In this chapter, you will compose a final map showing the result of the catchment delineation analysis. You will start with a basic print layout and enhance it with standard map elements. You will also learn some tricks such as use of variables and expressions to add some automation to the map.

After this chapter you will be able to:

- craft a basic map layout

- add standard map elements

- use a *variable* to add your name to the map

- use an *expression* to automate the map with the current date

- work with *map themes*

- export the map

7.2 Map Design Considerations - Adding More Data

In this chapter you will begin with the map document you saved at the conclusion of the "Stream and Catchment Delineation" chapter, on page 71. For a basemap you can choose either the OSM Standard basemap (via the QuickMapServices plugin), or the color hillshade basemap you created in the "Stream and Catchment Delineation" chapter (figure 7.1, on the following page). The examples that follow will use the color hillshade basemap.

While the data from the analysis are nicely styled, and the study area is highlighted using the *Inverted Polygon Shapeburst Fill* technique, it will be helpful to provide a little context for the map reader. While you have become familiar with the data during the analysis, a map reader needs some additional data to help understand the size and location of the catchment. When crafting a map it is very important to think about what information needs to be highlighted, who the audience will be, and how will the map be delivered.

For example, you will need to make very different cartographic decisions between a poster, an A4 (or letter), or even smaller peer-reviewed journal sheet sizes. Obviously the larger the sheet size, the larger the map scale, and the more detail you can include. Also think about whether it is a technical audience familiar with the data versus the general public? If it is a technical audience, then your work is a little easier. Maps for the general public require more explanation and finesse. Other concerns include whether the map can be produced in color, or if it needs to be black and white? Color gives you many more options. Producing small black and white maps for journals can be some of the hardest cartography! Also think about how the map will be delivered. Will it be printed or distributed digitally?

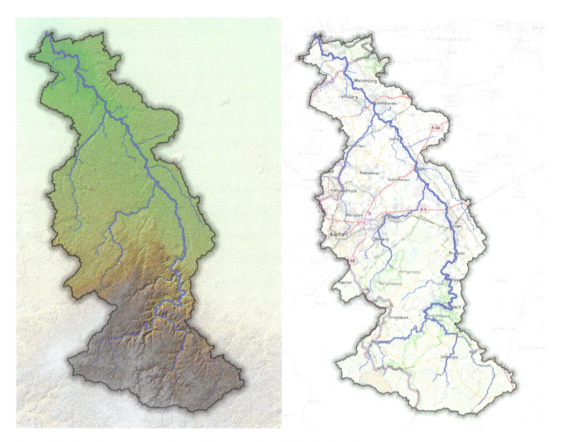

Figure 7.1: Rur Basin with a Color Hillshade (left) and OSM (right) basemap

You may occasionally need to produce a map for a color blind map reader. Fortunately QGIS gives you color blind previews. From the *View* menu choose *Preview mode.* Here you will be able to change your screen to simulate two types of color blindness (protanope & deuteranope) along with grayscale. The ColorBrewer 2.0 website (http://colorbrewer2.org) also gives you options for designing color ramps suitable for a color blind map reader.

For this exercise, you will assume a technical audience. You will produce a map on an A4 sheet in color. This will allow you to either print the map or share it digitally.

1. To begin, you will download cities and towns using the *QuickOSM* plugin. If you need to refresh your memory refer to section "Adding Vector Data from OpenStreetMap", on page 115 in the "Adding Open Data to your Catchment" chapter (page 111).

2. Use the *QuickOSM* plugin to download point data with the *Key* of *place*. For that *Key* you will search for and download *Values* of *city* and *town*.

3. You only need to download these data for the extent of the shaded relief.

4. Save these data into the the Rur_data.gpkg GeoPackage.

7.3 Styling and Labeling Cities and Towns

1. Open the *Layer Styling panel* by clicking the 🖌 button. Set the target layer to *Cities*.

2. They are styled by default with a *Simple marker*.

3. Select the *Marker* component and set the symbol to the *topo pop capital*. This is a symbol that installs with QGIS. It is composed of two different *Simple marker* symbols.

4. Next you will label the cities. Switch to the *Labels* tab ⟨abc of the *Layer Styling Panel*. Switch from *No Labels* to *Single labels*. Set the *Label with* option to the name field.

5. Change the *Font* to a san serif font such as *Arial* or *Calibri*. Labels tend to be small. San serif fonts are better suited for labels because they are simpler and are easier to read. Reduce the font *Size* to 9.

6. To emphasize these cities change the font *Style* to *Bold*.

7. Label buffers can be very effective at making labels easier to read, however, you never want the buffer to stand out. They are best when they enhance label readability but aren't initially noticeable. Switch to the *Label buffer* tab abc and check the *Draw text buffer* option. Set the *Color* to a light grey (RGB: 191|191|191). Next set the *Blending mode* to *Soft light*. Now there are subtle buffers that change with the background color. This article by Tom Armitage (http://learngis.uk/advanced-label-halos-in-qgis-3-x/) covers other ways to make halos effective against complicated backgrounds.

8. To give more separation between the labels and the feature icon switch to the *Label placement* ❖ tab and set the *Distance* to 2.5. Also change the *Placement* to *Cartographic*.

9. Next you will turn your attention to the *Towns*. Set the target layer of the *Layer Styling panel* to *Towns* and switch back to the *Symbology* tab.

10. Select the *Marker* component and set the symbol to *topo pop city*.

11. Switch to the *Labels* tab ⟨abc of the *Layer Styling Panel*. Switch from *No Labels* to *Single labels*. Set the *Label with* option to the name field.

12. Use the same font you used for *Cities* but reduce the size to 8 and keep the *Style* of *Regular*.

13. Switch to the *Label buffer* tab abc and use the same settings you used for *Cities*.

14. Switch to the *Label placement* ❖ tab and set the *Distance* to 2.

15. Finally for each point layer, click the *Automated placement settings* ❖ button to open the *Automated Placement Engine* window. Uncheck the box for *Allow truncated labels on edges of map* option. This will prevent labels from being cut off at the map borders.

16. At this point your map should resemble figure 7.2, on the following page.

Figure 7.2: Cities and Towns styled

7.4   Creating a Catchment Boundary Layer

1. Currently the Rur Catchment is styled as an Inverted Polygon Shapeburst Fill. Here you will duplicate the layer to create a version styled as a simple outline.

2. Right-click on the *Rur_ Catchment* layer and choose *Duplicate* from the context menu.

3. Turn the duplicated layer on, and move it above the Inverted Polygon Shapeburst Fill version.

4. Change the duplicated copy from an *Inverted Polygon* renderer to a *Single symbol*. Select the Shapeburst fill style and click the *Remove Symbol Layer* [⊟] button to delete it. Now there is a version of the boundary represented as a simple outline.

5. Expand *Layer Rendering* and increase the *Opacity* back to 100%.

7.5   Setting up the Print Layout

Now that you have provided a little context to the analysis results, you will create a new Print Layout and set up the page.

> Note that this task incorporates some changes to the layout environment introduced at QGIS versions 3.6 and 3.8. There is nothing that will prevent you from completing it with the 3.4 long-term release version. However, if using 3.4 you will see different options for setting the map extent in the Print Composer and adding North arrows.

1. First you will open a New Print Layout. There are three ways to do this.

   - From the menu bar choose `Project | New Print Layout`

   - Click the *New Print Layout*  button

   - Use the keyboard shortcut `Ctrl + P`.

2. Name the Layout `Rur Catchment` (figure 7.3).

Figure 7.3: Name the Print Layout

3. Click *OK*. A new Print Layout will open. This is where you craft your map.

The Print Layout is an application window with many tools that allow you to craft a map. For detailed information about the Print Layout, refer to the QGIS manual: `https://docs.qgis.org/3.4/en/docs/training_manual/map_composer/map_composer.html`. The main window of the Print Layout displays the piece of paper upon which the map will be designed. There are buttons along the left side of the window that allow you to add various map elements: map, scale bar, photo, text, shapes, attribute tables, etc. Note that each item added to the map canvas becomes a graphic object that can be moved, resized, and further manipulated (if selected) by the *Item Properties* tab on the right side of the layout. Across the top are buttons for exporting the composition, navigating within the composition, and some other graphic tools (grouping/ungrouping etc.) as shown in figure 7.4, on the following page.

4. To set the sheet size, right-click on the blank page and choose *Page Properties*. Here you can specify details about the overall composition. QGIS defaults to an A4 sheet size. There is no need to change that here since that is what you will use.

> QGIS Print Layouts allow you to add as many pages to a Layout as you wish for a given map document. These can also be of differing page sizes and orientations. They can even be in a different CRS from the main map canvas.

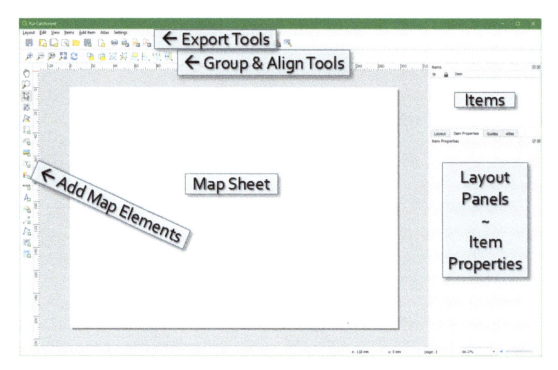

Figure 7.4: Print Layout Window

5. Set the *Orientation* to *Portrait*.

Note: Clicking the Save button in a Print Layout saves the entire map project.

7.6   Adding the Map

1. Using the *Add new map to layout* button, drag a box on the map sheet where you'd like the map to go. Remember that you'll need room for a title at the top of the page and a legend at the bottom of the map (figure 7.5, on the next page).

The map object can be moved and resized after its been added by selecting it with the *Select/Move item* tool. You can then use the handles around the perimeter to resize it. Remember when an object is selected, the *Item Properties* tab will show properties specific to that object.

2. The next step is to set the map extent within your composition. With the map selected click on the *Item Properties* tab. New to QGIS 3.6 are a series of buttons across the top of the panel for controlling the map extent.

From left to right these are:

- Update Map Preview
- Set Map Extent to Match Main Canvas Extent

- 🖼 View Current Map Extent in Main Canvas

- 🐎 Set Map Scale to Match Main Canvas Scale

- 🖼 Set Main Canvas to Match Current Map Scale

- 🖼 Interactively Edit Map Extent

- 🔖 Labeling Settings

> Note that these tools replace the *Set to map canvas extent* and *View extent in map canvas* buttons, found in all QGIS versions back to QGIS version 2.0.

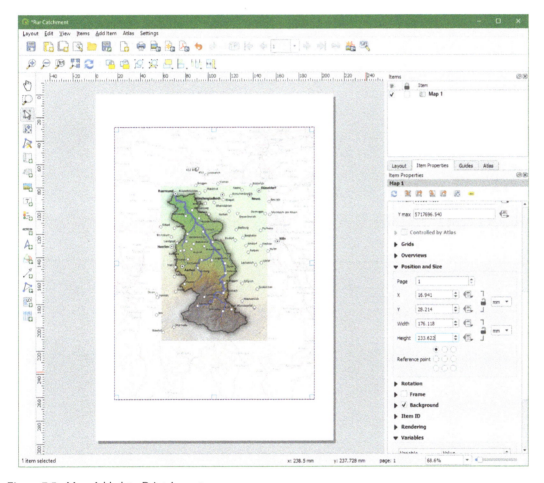

Figure 7.5: Map Added to Print Layout

3. Click the *Set Map Extent to Match Main Canvas Extent* 🖼 button. That will help orient the map on the sheet of paper as it appears in QGIS Desktop. However, if the aspect ratio of your QGIS Desktop map canvas doesn't match that of your print composition this may not give you the desired result.

If you need to make additional adjustments to the scale, you can do so in the *Main properties* section by adjusting the *Scale* value. Map scale is a ratio of Map Distance/Ground Distance. Here the number is roughly 750000 which can be read as a scale of 1:750,000. To zoom out you increase this number—reducing it zooms in. Clicking the *Update Map Preview* 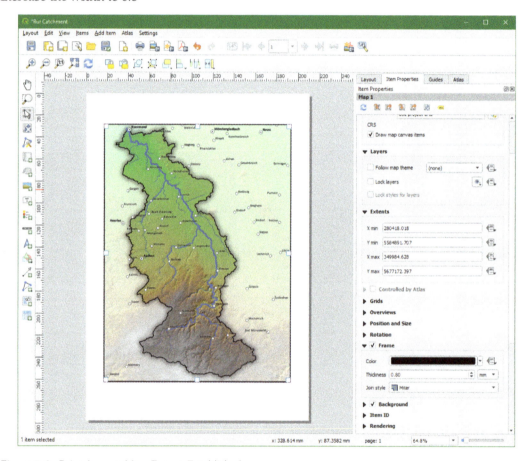 button forces the map view to refresh.

4. Set the *Scale* value to 395000 and hit the enter key. This should zoom the map to the boundaries of the Rur Catchment.

5. If you need to pan the map you can use the *Move Item Content* button. This allows you to pan the map content in the map frame without changing the scale. It is normal to have to make adjustments to get the map extent just right. Try to make your layout match figure 7.6.

6. Scroll down on the *Item Properties* tab and find the *Frame* section. Enable the *Frame* and increase the width to 0.8

Figure 7.6: Print Layout Map Extent Established

7.7   Adding a Title

Now you will add the title to your map. The purpose of a map title is to quickly convey the content and focus of the map to the reader. It should be concise and prominent.

1. Use the *Add new label* tool to drag a box all the way across the top of the composition above the map object. The text box can be resized after the fact by using the graphic handles.

2. By default the text box will be populated with the placeholder text *Lorem ipsum*. Using the

Item Properties tab *Main Properties* section, replace the holder text with the title: `Rur Catchment and Channels`.

3. In the *Appearance* section you can change the font. Click the *Font* button to open the *Text Format* window. Change the font to: Times New Roman, Bold, Size 36. You can use the search box above the font list to search for Times New Roman.

4. Below the font settings are some alignment settings. Set the Horizontal and Vertical alignment to center (figure 7.7).

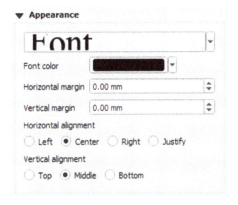

Figure 7.7: Title Font Settings

7.8   Adding a Legend

The purpose of the legend is to identify what symbols and colors on the map represent. Legends are used for data layers that are non-intuitive or require more explanation. For example, a point labeled *Rotterdam* does not need to be included in a legend, it is obvious what it is. However, the Strahler orders do need explanation.

1. Now you will add a legend. Use the *Add new legend* tool to drag a box on the lower-right corner beneath the map.

The legend will not fit in this space with the default font but you will change the settings. The only layers needed are the catchment and the channels. The *Item Properties* tab will be used to configure the legend (see figure 7.8, on the following page).

2. Uncheck *Auto update*. This will enable you to modify the legend, however, updates to the map will no longer be reflected in the legend unless you re-enable *Auto update*.

3. Select the Cities layer and click the *Remove item* button to remove it. Repeat to remove all the layers except the *RurChannels* and *Rur_catchment copy*.

4. Use the *Move Up* and *Move Down* buttons to move the *Rur_catchment copy* to the top of the legend.

5. Layers in the legend should not have original file names. They should be descriptive names that will be clear to the map reader. Now you will work on renaming the two layers. Select the

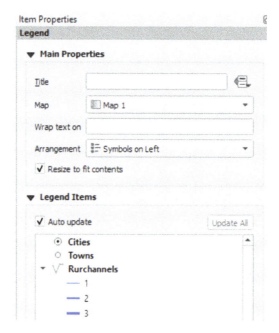

Figure 7.8: Legend Properties

RurChannels layer name and click the *Edit*  button. Change it so it reads *Rur Channels*.

6. Next select the values for the Strahler orders and rename each from 1,2,3, etc. to *Strahler Order 1, Strahler Order 2*.

7. Rename *Rur_catchment copy* to *Rur Catchment*.

8. Scroll down in the *Item Properties* panel to find the *Columns* section. Expand it. Set the *Count* to 2 and check *Split layers*. Your legend should now resemble figure 7.9).

Figure 7.9: Map Legend Configured

7.9   Adding a Scale Bar

Scale bars give the map reader a way to approximate distances on the map. There are two types: graphic scale bar or scale text. Here you will learn how to add a graphic scale bar.

1. Click on the *Add new scalebar* button.

2. Click to the left of the legend to add the scalebar to the map.

3. On the *Item Properties* tab, in the *Units* section make sure the *Scalebar units* are set to *Kilometers*.

4. In the *Segments* section, keep the left to 0 and the right to 2.

5. In the *Main Properties* section change the *Style* to *Line Ticks Down*. Notice that one of the styles is *Numeric* which would add scale as scale text (*1:100,000*).

6. In the *Fonts and Colors* section, reduce the *Font* to size 9.

7. Use the *Select/Move Item* ![cursor icon] tool to place the scalebar in a good position near the bottom of the white space in the lower left corner. Next you will be adding some descriptive text and a north arrow just above the scalebar, so leave space for those (figure 7.10).

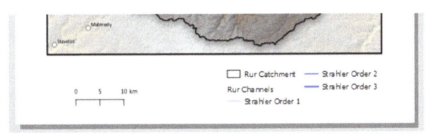

Figure 7.10: Scale Bar Added

7.10   Adding a North Arrow

Often it is nice to add a north arrow to a map composition to help orient the map reader. One should especially be added if north is not up on the map. Here you will learn how to add this to your map.

1. Click on the *Add north arrow* ![north arrow icon] button.

2. Drag a small box into the empty space just right of the scale bar. A default north arrow graphic will be added.

> Note that this Add north arrow tool was introduced at QGIS version 3.8. Prior to this you would have used the Add image tool. A north arrow can still be added that way if you prefer.

3. If you want to choose a different north arrow, find the *Search Directories* section on the *Item Properties* tab and expand it. A series of SVG graphics included with QGIS will appear. From here you can find those which are north arrows and choose a different one if you'd like.

4. Scroll down to the *Image Rotation* section of *Item Properties*. Note that *Sync with map* is enabled and is set to *Map 1* (figure 7.11, on the next page).

5. Scroll up to the *Main Properties* section and change the *Resize mode* to *Zoom and resize frame*.

6. Resize the north arrow graphic and move as needed so that it is well placed (figure 7.12, on the following page).

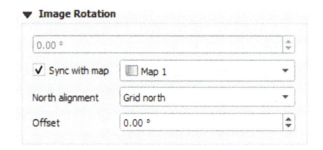

Figure 7.11: North Arrow Image Rotation Settings

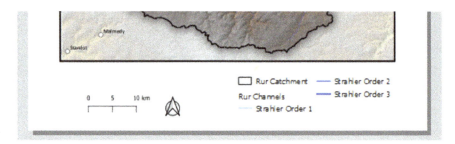

Figure 7.12: North Arrow Placed on Map Composition

7.11   Adding Descriptive Text

It is good practice to include credits for both data sources and cartography on a map. It can also be helpful to include details such as the date.

1. Next you will enter some descriptive text that tells the map reader where the data was obtained, who the cartographer was, and the date created. This will be done using the *Add new label* tool, the same tool you used to add the title. Add the label in the space above the scale bar and north arrow. Add the following text in the *Main properties* window of the *Item Properties* tab:

```
Data Sources:  SRTM and Natural Earth

Cartographer:  <your name>

Created on:  <todays date>
```

2. You can manually enter the date. However, it is also possible to use an expression to automate the date. To do this select the *Created on: <todays date>* text and click on the *Insert an expression* button.

3. The *Insert expression* window opens. Delete the *Created on: <todays date>* text.

4. Use the search box to find the String | concat() function.

- Double click on the function to enter it into the expression. This functions concatenates different blocks of text together.

- Inside the parenthesis type: 'Created on: '

- As you form your expression, you will see a small red triangle indicating an issue with your expression. This is to be expected since you haven't completed it yet. It is designed to be a helpful indicator of an invalid expression

- Enter a comma after this first text block (after the single quote).

- Now search for the `Conversions | to_date()` function.

- Double-click it to add it to your expression after the comma.

- Enter the variable `$now` as the single value for the `to_date()` function.

- End your expression with two closing parentheses.

5. Your expression should now look like:

`concat('Created on:  ', to_date($now))` (see figure 7.13)

- Click *OK* to insert your expression.

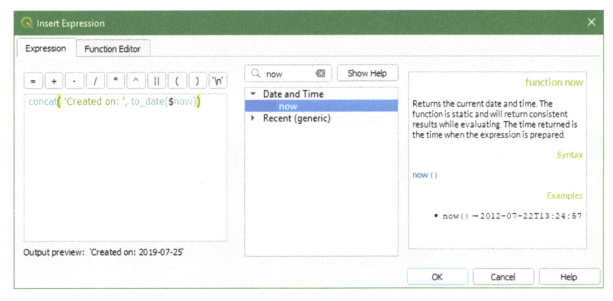

Figure 7.13: Date Expression Completed with Valid Output Preview

6. Format the expression so that it is just after the Cartographer line (figure 7.14).

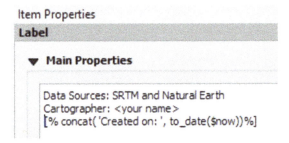

Figure 7.14: Layout Text with Date Expression

7. To finish this element, set the font to *size* of 8.

8. Configure the text element, scale bar, and north arrow so that they all fit in the space to the right of the legend. Note that with an element selected, you can also use the arrow keys on your keyboard to nudge.

> Did you know there is a recent enhancement to QGIS 3.6 that allows for custom layout checks? An example would be checks to warn a cartographer if desired map elements have been added or properly configured. These checks can help ensure maps meet a set of minimum organizational design criteria. Read more here: http://bit.ly/2G852H8

7.12   Using Variables for Adding Your Name as Author

QGIS allows you to store variables. This means any type of constant such as a unit conversion factor or your name. Variables can be set at several levels: Global, Project, and Layer. Here you will create a new Global variable with your name as cartographer.

1. Bring up the main QGIS desktop application.

2. From the menu bar choose Settings | Options.

3. Click on the ℰ Variables section. This is where *Global* variables are found.

4. Click the *Add variable* ⊕ button.

5. Replace *new_ variable* with *cartographer*.

6. Click in the *Value* cell to the right and type your name (figure 7.15). Click *OK*.

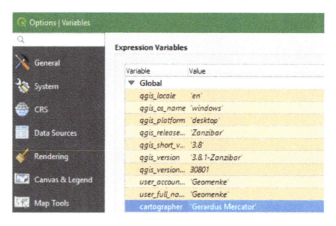

Figure 7.15: Creating a Cartographer Variable

7. Bring up the Print Layout window again.

8. Select the text you were most recently working with.

9. Highlight the name you entered for cartographer and click the *Insert an Expression* button.

10. Scroll down in the list of functions to the *Variables* section and expand it. Find the variable you just created named *cartographer*. Double-click on it to add it to your expression. When

you select it also notice that the value of that variable is displayed in the right hand pane of the expression window (figure 7.16).

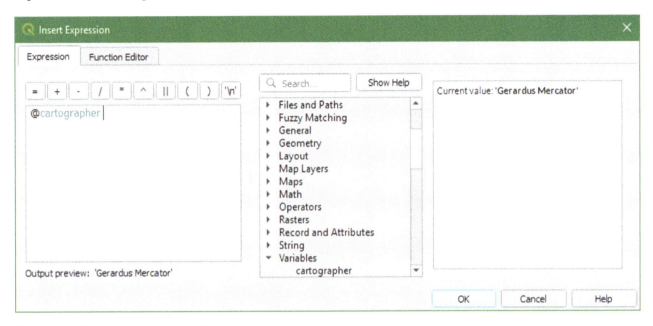

Figure 7.16: Using the Cartographer Variable

11. While you have the *Insert Expression* window open, notice that there is a *Recent* section. Expand it and you will find the date expression you created. You can use this *Recent* section to bring up recently used expressions and use them without having to recreate them!

12. Click *OK*.

13. The text in the *Items Properties* panel now reads: (figure 7.17).

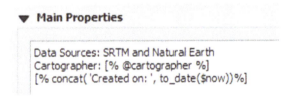

Figure 7.17: Text with both Date Expression and the Cartographer Variable

14. You have now used both an expression and a variable to automate your text. The @cartographer variable will always be available until you delete it. Variables and other expressions can be used throughout the QGIS interface to make your job easier! The date expression will update automatically each time the map is modified.

15. Your overall map should now resemble figure 7.18, on the following page.

7.13   Setting Up a Map Theme

In these final sections you will learn how to work with *map themes* to create a locator map. Locator maps are smaller than the main map. They have a smaller scale and provide a broader spatial context, showing where the main body of the map is located.

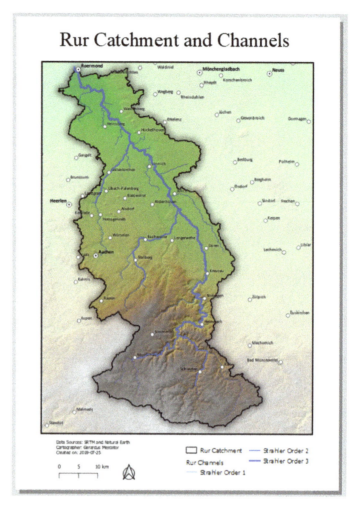

Figure 7.18: The Map Composition at this Point

This will involve:

- Creating a map theme from the current map and connecting the map layout to the theme

- Adding, styling, and labeling a country boundaries layer

- Creating a map theme for the locator map

- Adding a new map to the layout and connecting it to the locator map theme

- Configuring the Overview

1. Return to the QGIS Desktop window. Select all the layers in the *Layers panel* and choose *Group selected* from the context menu. Name the group *Main Map* (figure 7.19, on the next page).

2. Now you will set the current view as a map theme. Locate the *Manage Map Themes* drop down menu at the top of the *Layers Panel* and click on it. Choose *Add Theme* and the *Map Themes* window will open. Name the new theme *Main Map* and click *OK*.

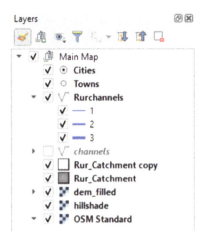

Figure 7.19: Main Map Layers Grouped

3. Return to the print layout window. Select the map object with the *Select/Move Items* tool. On the *Item Properties* tab find the *Layers* section. Click the box for *Follow map theme* and change it from *none* to *Main Map* (figure 7.20).

Figure 7.20: Setting the Map Object to Follow a Map Theme

7.14   Setting up the Layers for the Locator Map

Here you will add some data, set it up in a new Layer group and style it. The new Layer group will have two layers: Country boundaries and a copy of the Rur Catchment.

1. From the data included with the book find the MapComposition.gpkg. Add the CountryBoundaries layer found inside. If it lands within the existing *Main Map* group, right-click on it and choose *Move Out of Group*.

> This dataset was downloaded from Natural Earth (https://www.naturalearthdata.com). This is another fantastic public domain resource for global GIS data.

2. Click Ctrl + Shift + H to Hide All Layers. This is a keyboard shortcut that allows you to quickly turn off all the layers in a project. Turn the *CountryBoundaries* layer back on.

3. Find the copy of the Rur Catchment layer that is styled with a simple outline (not the one styled as an inverted polygon shapeburst fill). Right-click on it and choose *Duplicate*. Then right-click on the copy and choose *Move Out of Group*. You now have the two layers that will participate in the Locator map.

4. Select both these layers and choose *Group selected* from the context menu. Name the group *Locator Map* (figure 7.21).

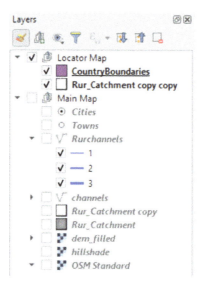

Figure 7.21: Both the Main Map and Locator Map Layer Groups Configured

5. Style the CountryBoundaries with a *Fill style* of *No brush*. Keep the default *Stroke color* of black but increase the *Stroke width* to 0.46.

6. Switch to the *Labels* tab (abc) of the *Layer Styling Panel*. Label the CountryBoundaries with the NAME field.

7. Change the *Font* to a san serif font such as *Arial* or *Calibri*. Reduce the font *Size* to 9 and change the font *Style* to *Bold*.

8. Now you will style the Rur Catchment for the Locator map. Select the *Simple line* component. Change the *Symbol layer type* to *Simple fill*.

9. Change both the *Fill color* and *Stroke color* to RGB: 0|145|255 (figure 7.22, on the facing page).

10. Now you will set the current view as the Locator map theme. From the *Manage Map Themes* drop down menu again choose *Add Theme*. In the *Map Themes* window name the new theme *Locator Map* and click *OK*.

11. Return to the print layout window. You will now add a second map to the layout.

12. Using the *Add new map to layout* button, drag a small box on the existing map object along the upper right side, being careful to not cover the Rur Catchment.

13. You will now set the *Map theme* for this second map. On the *Item Properties* tab, find the *Layers* section. Click the box for *Follow map theme* and change it from *none* to *Locator Map* (figure 7.23, on the next page).

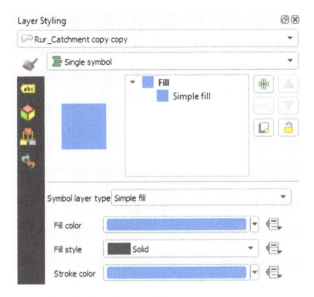

Figure 7.22: Rur Catchment Styling for the Locator Map

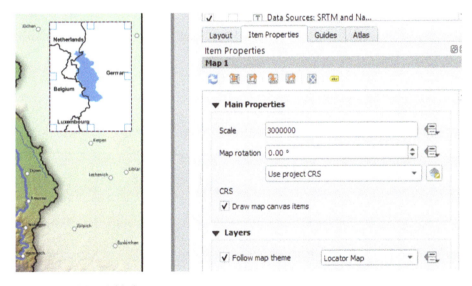

Figure 7.23: Locator Map Added

> On the Item Properties tab for map objects, there is also an *Overviews* section which allows you to configure a box indicating for example, the spatial envelope of the main map canvas within the locator map. In this case you did not use that, because including the watershed on the locator is sufficient. However, it is often useful to include *Overviews* with inset and locator maps.

7.15   Adding a Gradient Legend for Elevation

This entire analysis was based on the elevation layer. The bad news is that QGIS doesn't support creating a gradient legend via the standard Legend item. The good news is that it can be done by adding a rectangle, and filling it with the same color ramp used for the elevation. Then you simply add the text labels. With the current layout you have room along the lower left corner of the map where this second legend will fit nicely.

1. Click the *Add shape* button and select the ☐ **Add Rectangle** tool. Use it to drag a long narrow rectangle in the lower-left corner. On the *Item Properties* tab expand the *Position and Size* section and give this a *Width* of 5 and *Height* of 50 (figure 7.24).

Figure 7.24: Rectangle for Gradient Legend Added

2. Now under *Main properties* click the white color bar for the *Style*. Select the *Simple fill* component. Change the *Symbol layer type* to *Gradient fill*.

3. Choose the *Color ramp* option. Go through the same steps used previously to choose the same color ramp used in the dem_filled.

- *Create New Color Ramp*

- Type = *Catalog: cpt-city*

- *Topography - sd-a*

4. The legend will look nicer with a black border. Click the *Add symbol layer* ⊞ button. Select the *Simple fill* component and set the *Fill style* to *No Brush* (figure 7.25, on the next page).

5. Now use the *Add new label* T tool, to add labels for the upper and lower elevation values to the right of the gradient legend. Also add the word *Elevation* just above the gradient legend (figure 7.26, on the facing page).

> On this map the elevation is being blended with the hillshade layer using the Multiply blending mode. This darkens the colors a bit. To more truly represent what is being seen on the map you can copy and paste this rectangle and align the copy with the original. Give this upper rectangle a *Simple fill* with a *Color* of gray (70%) and an *Opacity* of 30%.

7.16 Final Adjustments

1. Return to the main QGIS Desktop window. Give the labels for the CountryBoundaries a default white *Buffer*. This will help the labels be more readable where they cross national boundaries.

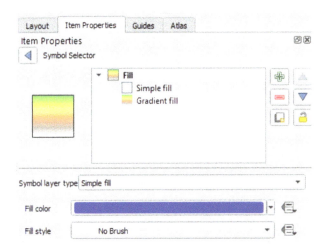

Figure 7.25: Gradient Legend Styling

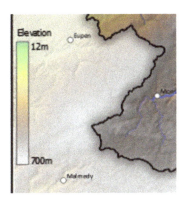

Figure 7.26: Gradient Legend

2. Return to the print layout window and click the *Refresh map* button to see the change.

3. Make sure the Locator Map is still selected. From the *Item Properties* window, decrease the scale (zoom out) of the Locator map to about 3000000 so the labels better fit within the national boundaries.

4. Scroll down and activate the *Frame* for the Locator map.

5. It may be that the Locator map is interfering with City and Town labels on the main map. To avoid this conflict, select the Main map object. Near the top of *Item Properties* find the *Labeling Settings* button.

6. This opens the *Label settings* panel. Here you can select map elements as label blocking elements (figure 7.27, on the next page). Click *Map 2*. Now the labels for Cities and Towns on the main map will move to avoid conflicts with the Locator map. This can be done for any the map elements.

Did you know that you can also move labels to custom positions? For example, you may want to shift some country labels in the Locator map. From the main QGIS Desktop window enable the Labels toolbar. Select the layer with the labels you wish to move. Find the button on the toolbar named *Move Label and Diagram*. Click on a label. You will be prompted to identify the unique ID column of the layer. Once that has been done you can click on individual labels and move them to custom locations. On the same toolbar there are also tools for Rotating labels and altering fonts (Change label).

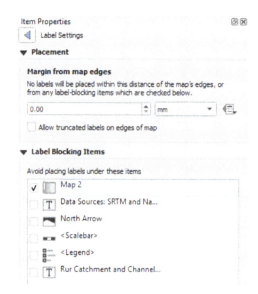

Figure 7.27: The Locator Map Configured as a Label Blocking Item

7.17   Exporting the Map

1. Congratulations your map is finished! The final step is to export it to a high-resolution image.

2. Click the *Export as image* button.

3. The *Save Layout As* window opens.

- Navigate to the exercise folder.

- Choose a *Save as* type such as JPG or PNG.

- Name the file and click *Save*.

- The *Image Export Options* window will open. Keep the default 300 dpi setting and click *Save*.

Note that maps can also be exported as a PDF or SVG file.

4. Once the map has finished exporting, a link to the folder will appear across the top of your print layout. You can click on the link to open your systems file browser and see the result (figure 7.28).

5. The final map should look like figure 7.29, on the next page.

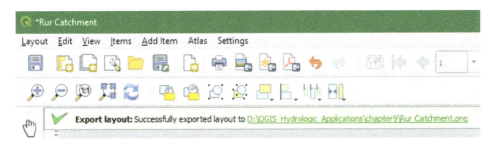

Figure 7.28: Map Export Link

Rur Catchment and Channels

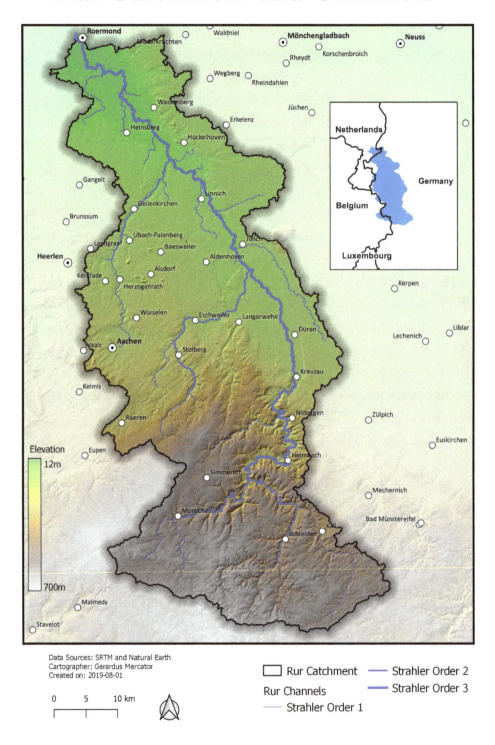

Figure 7.29: Final Map

8. Conclusion

Having reached this chapter, you've made it through the exercises of QGIS for Hydrological Applications. We hope you've enjoyed learning useful features and plugins that QGIS offers for catchment hydrology and water management. Now it is important to apply the skills to your own cases and add different flavors to the recipes presented in this book.

It was hard to make a selection of exercises for the book. QGIS has many more useful features and plugins to explore. Besides QGIS there are many other FOSS4G tools that can be used in hydrological applications, such as PostGIS, GDAL/OGR, and many Python libraries, which are covered by other books from Locate Press.

On the IHE Delft OpenCourseWare site for GIS (http://www.gisopencourseware.org) you can find frequently updated learning resources for hydrological applications, including GDAL and Python tutorials. In addition, the videos on the YouTube channel (https://www.youtube.com/c/HansvanderKwast) show the latest developments.

There is a new QGIS release every four months and a new long-term release annually. Stay current on new features by browsing the Visual Changelogs at https://www.qgis.org/en/site/forusers/visualchangelogs.html

8.1 Things to Do

Learn more skills via the QGIS Training Manual: https://www.qgis.org/en/site/forusers/trainingmaterial/index.html.

Refer to the QGIS User Guide when technical questions arise: https://docs.qgis.org/3.4/en/docs/user_manual/.

Look on GIS.StackExchange.com for solutions to issues you encounter: https://gis.stackexchange.com/questions/tagged/qgis.

Subscribe to a QGIS Mailing List: https://www.qgis.org/en/site/getinvolved/mailinglists.html.

Read about, How to ask a QGIS question before posing your first question to a mailing list (list-serv): https://www.qgis.org/en/site/getinvolved/faq/index.html#how-to-ask-a-qgis-question.

Read about how others have used QGIS in QGIS Case Studies: https://www.qgis.org/en/site/about/case_studies/index.html.

Explore maps others have made using QGIS in the Flickr Map Showcase: https://www.flickr.com/groups/qgis/pool/.

Read blog entries about new features and how-tos at http://plugins.qgis.org/planet/:

- Anita Graser: https://anitagraser.com/
- North Road: https://north-road.com/blog/

- Klas Karlsson: `https://www.youtube.com/channel/UCxs7cfMwzgGZhtUuwhny4-Q`

- Spatial Thoughts: `https://spatialthoughts.com/blog/`

Explore other QGIS Applications such as QGIS Server and QGIS Web Client: `https://www.qgis.org/en/site/about/features.html`

If you require commercial support for QGIS you can go to this page to see companies providing a variety of support options: `https://www.qgis.org/en/site/forusers/commercial_support.html`.

We also hope that this book has inspired you to get involved as a QGIS community member. There are different ways to contribute to QGIS. You can help reporting bugs, translating QGIS, develop plugins, develop new features, and sponsor/donate to QGIS. Read more here: `https://www.qgis.org/en/site/getinvolved/index.html`.

Join your local user group or start one if one doesn't exist. Visit this page to see where there are established user groups: `https://www.qgis.org/en/site/forusers/usergroups.html`.

Visit the Open Source GeoSpatial Foundation (OSGeo) website, to stay abreast of other FOSS4G project news and conference announcements: `http://www.osgeo.org/`.

Index

Books from Locate Press

Be sure to visit http://locatepress.com for information on new and upcoming titles.

Discover QGIS 3.x

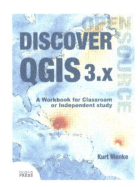

EXPLORE THE LATEST LONG TERM RELEASE (LTR) OF QGIS!
Discover QGIS 3.x is a comprehensive up-to-date workbook built for both the classroom and professionals looking to build their skills.

Designed to take advantage of the latest QGIS features, this book will guide you in improving your maps and analysis.

Discover QGIS 3.x is an update of the original title, using QGIS 3.6, covering Spatial analysis, Data management, and Cartography. The book includes new exercises and a new section—Advanced Data Visualization.

The book is a complete resource and includes: lab exercises, challenge exercises, all data, discussion questions, and solutions.

Leaflet Cookbook

COOK UP DYNAMIC WEB MAPS USING THE RECIPES IN THE LEAFLET COOKBOOK.
Leaflet Cookbook will guide you in getting started with Leaflet, the leading open-source JavaScript library for creating interactive maps. You'll move swiftly along from the basics to creating interesting and dynamic web maps.

Even if you aren't an HTML/CSS wizard, this book will get you up to speed in creating dynamic and sophisticated web maps. With sample code and complete examples, you'll find it easy to create your own maps in no time.

A download package containing all the code and data used in the book is available so you can follow along as well as use the code as a starting point for your own web maps.

QGIS Map Design - 2nd Edition

LEARN HOW TO USE QGIS 3 TO TAKE YOUR CARTOGRAPHIC PRODUCTS TO THE HIGHEST LEVEL.
QGIS 3.4 opens up exciting new possibilities for creating beautiful and compelling maps!

Building on the first edition, the authors take you step-by-step through the process of using the latest map design tools and techniques in QGIS 3. With numerous new map designs and completely overhauled workflows, this second edition brings you up to speed with current cartographic technology and trends.

See how QGIS continues to surpass the cartographic capabilities of other geoware available today with its data-driven overrides, flexible expression functions, multitudinous color tools, blend modes, and atlasing capabilities. A prior familiarity with basic QGIS capabilities is assumed. All example data and project files are included.

Get ready to launch into the next generation of map design!

The PyQGIS Programmer's Guide

WELCOME TO THE WORLD OF PYQGIS, THE BLENDING OF QGIS AND PYTHON TO EXTEND AND ENHANCE YOUR OPEN SOURCE GIS TOOLBOX.

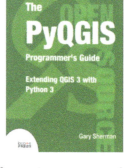

With PyQGIS you can write scripts and plugins to implement new features and perform automated tasks.

This book is updated to work with the next generation of QGIS—version 3.x. After a brief introduction to Python 3, you'll learn how to understand the QGIS Application Programmer Interface (API), write scripts, and build a plugin.

The book is designed to allow you to work through the examples as you go along. At the end of each chapter you will find a set of exercises you can do to enhance your learning experience.

The PyQGIS Programmer's Guide is compatible with the version 3.0 API released with QGIS 3.x and will work for the entire 3.x series of releases.

pgRouting: A Practical Guide

WHAT IS PGROUTING?

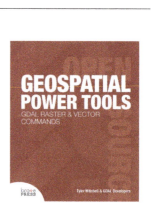

It's a PostgreSQL extension for developing network routing applications and doing graph analysis.

Interested in pgRouting? If so, chances are you already use PostGIS, the spatial extender for the PostgreSQL database management system.

So when you've got PostGIS, why do you need pgRouting? PostGIS is a great tool for molding geometries and doing proximity analysis, however it falls short when your proximity analysis involves constrained paths such as driving along a road or biking along defined paths.

This book will both get you started with pgRouting and guide you into routing, data fixing and costs, as well as using with QGIS and web applications.

Geospatial Power Tools

EVERYONE LOVES POWER TOOLS!

The GDAL and OGR apps are the power tools of the GIS world—best of all, they're free.

The utilities include tools for examining, converting, transforming, building, and analysing data. This book is a collection of the GDAL and OGR documentation, but also includes new content designed to help guide you in using the utilities to solve your current data problems.

Inside you'll find a quick reference for looking up the right syntax and example usage quickly. The book is divided into three parts: *Workflows and examples*, *GDAL raster utilities*, and *OGR vector utilities*.

Once you get a taste of the power the GDAL/OGR suite provides, you'll wonder how you ever got along without them.